AF845716

MICROBIAL PHYSIOLOGY

MICROBIAL PHYSIOLOGY

Edited by

Dr. Esha Rami

Delve
PUBLISHING

www.delvepublishing.com

Microbial Physiology

Dr. Esha Rami

Delve Publishing
4164 Lakeshore Road
Burlington, ON L7L 1A4
Canada
www.delvepublishing.com

Email: orders@arclereducation.com

© 2025 Delve Publishing

ISBN: 978-1-77956-365-1 (Hardcover)

This book contains information obtained from highly regarded resources. Reprinted material sources are indicated and copyright remains with the original owners. Copyright for images and other graphics remains with the original owners as indicated. A Wide variety of references are listed. Reasonable efforts have been made to publish reliable data. Authors or Editors or Publishers are not responsible for the accuracy of the information in the published chapters or consequences of their use. The publisher assumes no responsibility for any damage or grievance to the persons or property arising out of the use of any materials, instructions, methods or thoughts in the book. The authors or editors and the publisher have attempted to trace the copyright holders of all material reproduced in this publication and apologize to copyright holders if permission has not been obtained. If any copyright holder has not been acknowledged, please write to us so we may rectify.

Notice: Registered trademark of products or corporate names are used only for explanation and identification without intent of infringement.

Delve Publishing publishes wide variety of books and eBooks. For more information about Delve Publishing and its products, visit our website at www.delvepublishing.com.

ABOUT THE EDITOR

Dr. Esha Rami is presently working as an Assistant Professor, in the Department of Life science, Parul Institute of Applied Science, Parul University, India. She Did her Post graduated from Ganpat University, Ph.D. in biotechnology from Hemchandracharya North Gujarat University, in the year 2015. She has authored a number of national and international publications in reputed journals.

TABLE OF CONTENTS

LIST OF FIGURES

PREFACE

Our understanding of prokaryote functions and activities in the environment, including their symbiotic and pathogenic interactions, as well as their use in biotechnology, is based on our knowledge of their physiology and metabolism. Prokaryotic species (bacteria and archaea are characterized by amazing diversity in their metabolism and physiology, despite maintaining a relatively tiny size and simple structure throughout their evolutionary history. In contrast to eukaryotes, which show a limited or restricted metabolic variety, prokaryotes may flourish in a wide range of habitats and often dominate the biota. This is reflected in their metabolic diversity. As a result, prokaryotes are present practically everywhere and in a variety of environmental conditions, such as extremes of acidity and alkalinity, light and dark, low and high pressure, low and high salt concentrations, and availability/unavailability of nutrients. Certain bacterial families/species possess unique physiologies, such as nitrogen fixation, while other prokaryotic abilities include the utilization of inorganic substances as electron acceptors during respiration. Agroecosystem productivity is impacted by biotic and abiotic stressors. The use of chemicals to increase agricultural yield has made the issue worse. The inappropriate practices have caused the agroecosystems to become stressed in addition to becoming more and more polluted.

Fertilizers and pesticides have damaged people and other living things and turned the soil salinized, contaminated water supplies, and reduced biodiversity. Uncontrolled use of hazardous pesticides in crop fields has created problems for several beneficial bacteria found in plants. Even though these chemicals caused the green revolution, agriculturists are now aware of the negative consequences of the chemically induced green revolution. The issue has gotten worse because of the effects of climate change brought on by industry, urbanization, and population growth. Among the most important concerns of the modern era is ensuring food and nutritional security. Maintaining the diversity and viability of the agroecosystems makes the task even more difficult. Using more environmentally friendly farming practices and reducing the usage of dangerous pesticides is the possible solution.

This book can be used as a guide to undergraduate students with an interest in studying microbes.

INTRODUCTION

Nearly all plant species have symbiotic relationships with beneficial bacteria that can be used to increase the quantity and quality of food produced. The only things that can replace the hazardous chemicals are these microorganisms, also referred to as plant growth-promoting bacteria. Globally, there is already encouragement for the use of biopesticides and fertilizers. To effectively use these helpful bacteria in boosting productivity and preserving the health of the plant, we still need to comprehend the mechanisms using the most recent biotechnological methods. Researchers in the fields of agriculture microbiology and biotechnology are aware that using these plant-beneficial bacteria will bring about the next green revolution, which will provide a nutritious and balanced food while also preserving the sustainability of agroecosystems. Beneficial microorganisms for plants interact with the plant, assisting it in surviving and strengthening its defenses against biotic and abiotic stresses. These bacteria support the plant by supplying nutrients and killing phytopathogens through a variety of strategies. To utilize the plant microbiome for nutrition, and defense against abiotic stresses like salinity, temperature, and drought, it is vital to comprehend the mechanisms underlying the interactions between the plant and its microbiome. Plant and microbial nutrition have become a recognized field in plant physiology mostly due to the economic advantages of an efficient fertilizer schedule for an agricultural or horticultural crop.

This book provides basic knowledge of microbial physiology for all students.

Chapter 1

INTRODUCTION TO MICROBIAL PHYSIOLOGY

Contents

1.1 INTRODUCTION

The study of many distinct microorganisms is the scope of the vast field of microbial physiology. Bacteria are classified into spherical (coccus), rod (bacillus), and spiral (spirillum,, and comma shaped (Pavlova et al., 2022). Bacteria is a prokaryotic, unicellular organism. Since bacteria are prokaryotic, they do not have organelles such as a nucleus in their cells.. Apart from these fundamental types of bacteria, there exist highly specialized forms known as mycelial, budding, and sheathed forms. The cell surface serves as the microbial cell interface with the outside. It preserves the integrity of the cell as a distinct entity and shields the interior of the cell from outside threats/risks. It must also permit the movement of large molecules into and out of the cell. These molecules include proteins exported to the outside of the cell, vitamins (vitamin B_{12}), amino acids, nucleosides, and carbohydrates (glucose). Depending on the organism, distinct cell surfaces can have varied compositions and structures. Christian Gram, a Danish researcher, created a differential stain in 1884 based on the capacity of some bacterial cells to hold/retain the dye crystal violet following 95% ethanol decolorization(Moyes et al., 2009).

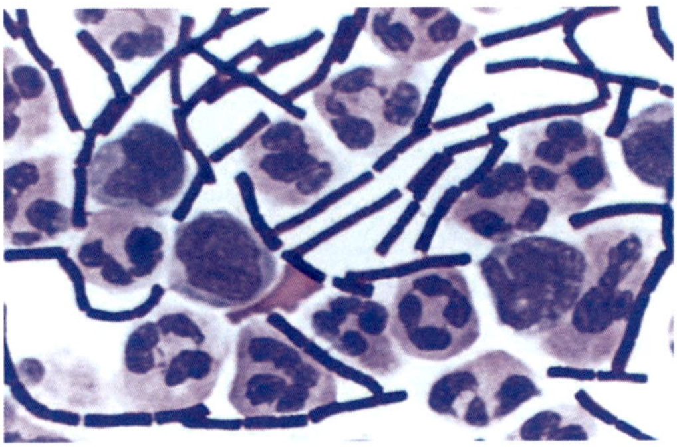

Figure 1.1 Gram-positive bacteria

Source: https://upload.wikimedia.org/wikipedia/commons/7/79/Gram_Stain_Anthrax.jpg,*Creative Commons License.*

Gram-positive cells (Figure 1.1) were defined as those that retain the stain. Compared to gram-positive cells, gram-negative cells have a more complicated surface. The cell wall and membrane are the two main structural components of the gram-positive cell surface. Multiple layers of peptidoglycan, a linear polymer made up of units of N-acetylglucosamine (NAG) and N-acetylmuramic acid (NAM), make up the cell wall of gram-positive bacteria. Muramic acid has a short peptide chain connected to it. The peptide chains in bacterial cell walls frequently link to one another. The cross-bridging between neighboring peptides in a gram-positive bacterium like *Staphylococcus aureus* (Figure 1.2) may be almost 100% (Lang et al., 2023). In contrast, *Escherichia coli*, (Figure 1.3) a gram-negative bacterium, may exhibit as little as 30% cross-bridging frequency. Other components are generated at the membrane surface and extend through the peptidoglycan layer to the outer surface. One such component is lipoteichoic acid, which is only found in gram-positive organisms/species.

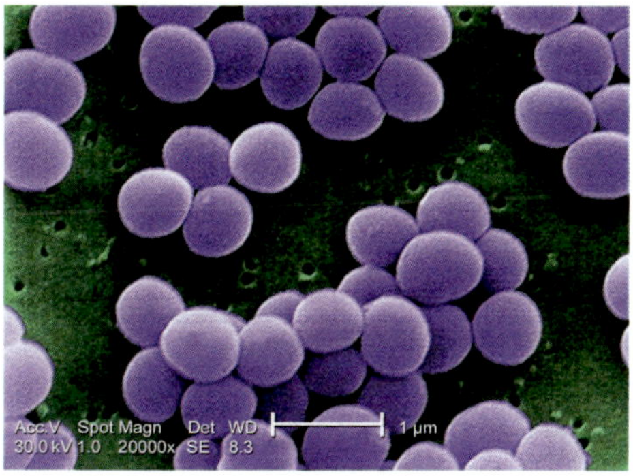

Figure 1.2 *Staphylococcus aureus*

Source: https://upload.wikimedia.org/wikipedia/commons/d/d3/Staphylococcus_aureus_VISA_2.jpg,*Creative Commons License.*

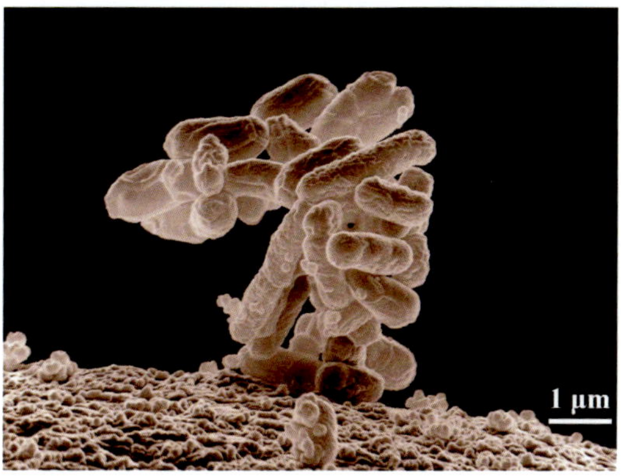

Figure 1.3 *Escherichia coli*

Source: https://upload.wikimedia.org/wikipedia/commons/b/bc/E_coli_at_10000x%2C_original.jpg,*Creative Commons License.*

1.2 CELL MEMBRANE

A gram-negative cell peptidoglycan layer typically consists of a single monolayer. The gram-negative cell is surrounded by an outer membrane made of lipoproteins, phospholipids, lipopolysaccharides, and other proteins. The term periplasmic space refers to the area between this external membrane and the inner membrane. Different enzymes and other proteins may pass through it multiple times. Both gram-positive and gram-negative cells have a lipid bilayer made up of various proteins, glycolipids, and phospholipids in their cytoplasmic membranes. While some of these proteins help the membrane structurally, others are involved in the movement of metabolites such as glucose and amino acids. Gram-negative cells have a relatively high lipopolysaccharide concentration in their outer membrane. These lipid-containing elements make up the O antigens, which are generated by the lipopolysaccharide exterior polysaccharide chains and are one of the key characteristics that distinguish gram-negative cells from other types. This lipid-containing material also exhibits endotoxin activity, in cases of severe infections brought on by gram-negative bacteria(Seydel et al., 1999). Additionally, certain protein or carbohydrate receptor sites on bacterial cell surfaces are available for the adhesion of bacteriophages, which are viruses that infect bacteria. After adhering to these receptor sites, the bacteriophage can commence cell invasion.

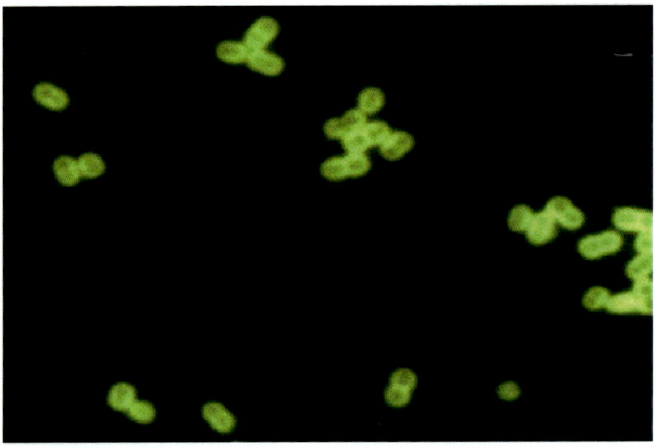

Figure 1.4 *Streptococcus pneumoniae*

Source: https://upload.wikimedia.org/wikipedia/commons/9/9f/Pneumococcus_CDC_PHIL_ID1003.jpg,*Creative Commons License.*

The paths/methods used by gram-positive and gram-negative cells to transport materials outside the membrane and inside the cell are slightly different. Gram-positive organisms have instant access to medium components through their cytoplasmic membrane. But before reaching the cytoplasmic membrane, chemicals, and nutrients in gram-negative bacteria must first pass through the outer membrane. Large molecules can enter the periplasmic space through gaps in the outer membrane of gram-negative cells. In gram-positive and gram-negative cells, subsequent transport across the inner or cytoplasmic membrane is similar. Certain bacterial cells secrete a capsule or slime layer that surrounds the cell. Polysaccharides are large molecular weight polymers of carbohydrates, or polypeptides, are polymers of amino acids that are frequently produced from the D-isomer of an amino acid. The β-1, 3- and β-1, 4-linked glucose and glucuronic acid make up the *Streptococcus pneumoniae* (type III) (Figure 1.4) capsule(Xiong et al., 2018).

The pneumococcus pathogenicity is caused by its capsular polysaccharide, also known as pneumococcal polysaccharide. One of the virulence factors of the anthrax bacillus, *Bacillus anthracis*, is the production of a polypeptide capsule made of D-glutamic acid subunits. Many microorganisms can move in an aqueous environment. In the case of bacteria, this mobility is achieved by the use of protein threads termed flagella, which are helical organelles. A basal body attaches the bacterial flagellum to the cell surface. The motor in the basal body rotates the flagellum, allowing the organism to move through the liquid/aquatic environment. Compared to flagella, several bacteria have shorter, more rigid exterior structures. These formations are known by the terms fimbriae or pili. Additionally, it appears that the source of these appendages is a basal body that is either inside the cytoplasmic membrane or in the cytoplasm below it. Cellular attachment to surfaces or host cells is facilitated by either generalized or common pilus.

1.3 COMPOSITION

Many electron micrographs show the tiny granular texture of the cytoplasm, which is present in all cells. This appearance is caused by minute particles known as ribosomes. About 35% of proteins and 65% of RNA make up ribosomes. Protein synthesis, or the polymerization of amino acids into proteins, is coordinated by the ribosome. The ribosome particles appear spherical when viewed under an electron microscope.

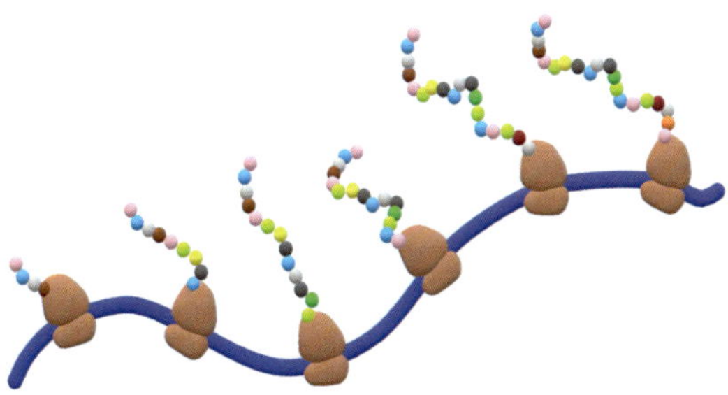

Figure 1.5 Polyribosomes

Source: https://upload.wikimedia.org/wikipedia/commons/a/a4/Multiple_Ribosomes_Translation_Protein_Synthesis.png,*Creative Commons License.*

Polyribosomes,(Figure 1.5)alsoknownaspolysomes,aregroupsofribosomes that are detected on a single messenger RNA (mRNA) molecule(Larkins, 1985). Using sucrose gradient sedimentation, it was discovered that the roughly spherical ribosome particle had a Svedberg coefficient of 70S. In the production of polypeptides, only the entire 70S is active. In contrast, eukaryotic cells have larger (80S) ribosomes made up of 40S and 60S subunits that are connected to the endoplasmic reticulum(Panse & Johnson, 2010).

Both 70S and 80S ribosomes have the same purpose in the production of proteins. It is interesting to note that eukaryotic mitochondria have 70S ribosomes instead of the expected 80S particles. This is because mitochondria most likely originated from endosymbiotic prokaryotic cells, a theory that has been extensively confirmed by genome comparisons between bacteria and mitochondria. The single, circular, double-stranded DNA molecule known as the *E. coli* chromosome has all the information needed for cell growth and structure encoded in its nucleotide sequence. Starting with the chromosome, the main molecular processes needed for the species to proliferate are transcription, translation, and DNA replication. In bacteria, replication entails the precise duplication of chromosomal DNA and the binary fission that creates two daughter cells. When a cell reaches a specific mass-to-DNA ratio during binary fission, fresh DNA is generated and a cross-wall is built in the center, which will eventually divide the two daughter cells.

1.4 GENETICS OF CELLS

The origin serves as the starting place for the unwinding of the double-stranded DNA molecule. On the other side of each strand, fresh DNA is created. DNA polymerase, the replication-related enzyme, positions adenine residues opposite thymine and cytosine residues opposite guanine using a parent strand as a template. From the origin, new DNA is synthesized in both directions until the replication forks meet at the terminus 180° from the origin. Cell division is now underway, with the creation of a cross-wall between the two freshly manufactured chromosomes. As the daughter chromosomes split apart, we can observe that the chromosome is linked to the cell membrane. Before the cell finishes building the cross-wall that will divide the two progeny, chromosomal segregation into pre-daughter cells must take place. Some processes are involved in the processing of the genetic information found in DNA to create different proteins. The start of a gene (promoter) is where the enzyme RNA polymerase (DNA-dependent RNA polymerase) is found(Chamberlin, 1982). Following a localized unwinding of a section of the chromosome, RNA polymerase transcribe RNA from the DNA template. The two subunits of the ribosome, 30S and 50S, are made up of ribosomal ribonucleic acids (rRNA) and unique ribosomal proteins. While rRNA molecules serve as the structural framework that guides the assembly of proteins to form a ribosome, they do not, by themselves, code for any proteins. The ribosome reads three nucleotides, referred to as a triplet codon, as a particular amino acid to convert mRNA into protein. Before being utilized by the ribosome, each amino acid (AA) needs to be linked to a transfer RNA (tRNA) molecule that is designed specifically for that amino acid. A charged tRNA molecule is defined as a tRNA with an attached amino acid. The codon in mRNA and the anticodon, a portion of the tRNA molecule, will form a base pair.

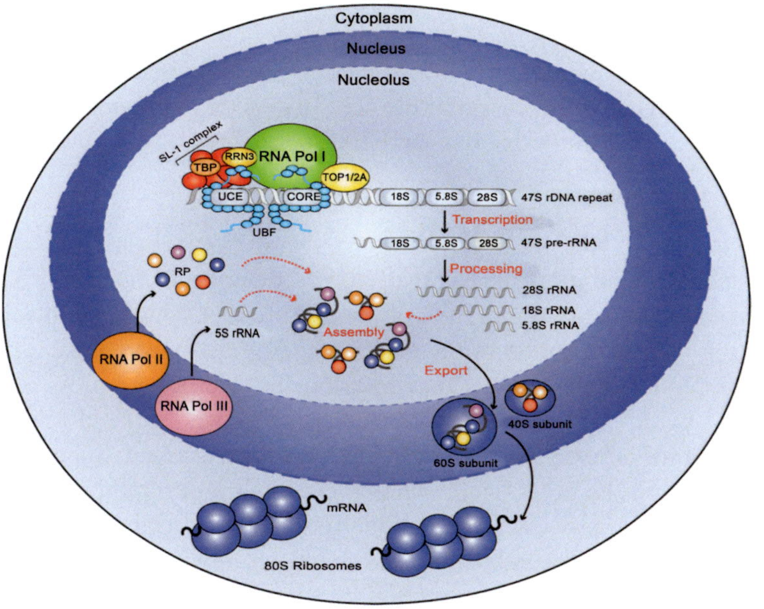

Figure 1.6 RNA polymerase

Source: Ferreira, R.; Schneekloth, J.S., Jr.; Panov, K.I.; Hannan, K.M.; Hannan, R.D. Targeting the RNA Polymerase I Transcription for Cancer Therapy Comes of Age. Cells 2020, 9, 266. https://doi.org/10.3390/cells9020266,*Creative Commons License.*

A peptide bond between 2 amino acids is catalyzed by the ribosome when two of these charged tRNA molecules occupy neighboring locations on the ribosome at the same time(Spirin, 2002). At this stage, one tRNA has two amino acids bound to it, while the other tRNA is uncharged and will soon be freed from the ribosome. After that, the ribosome is free to carry the message to the subsequent codon. A full protein is created when the ribosome reaches the end of the message. The N-terminal amino acid must initiate protein synthesis, whereas the C-terminal amino acid completes the process.

Furthermore, RNA polymerase (Figure 1.6) reads the DNA strand encoding the mRNA beginning at the 3' end, whereas the ribosome starts translating at the 5' end of the mRNA. The strand that essentially acts as an RNA polymerase template is not referred to by the term 5' end, which is typically used to describe the start of a gene. It expresses the complementary DNA strand that shares the same sequence as the mRNA with the exception that T is present rather than U. All of the fundamental building blocks and the macromolecules made from them must be created in precise amounts for a cell to grow effectively. It is critical to comprehend how a microbial cell controls the synthesis and concentration of each product in complex metabolic pathways. Feedback inhibition of enzyme activity is one of the common mechanisms of both genetic and metabolic regulation. In feedback inhibition, the reaction product inhibits the activity of an enzyme that is already present in the cell. In genetic repression, the reaction product inhibits the synthesis of an enzyme.

1.5 ENZYME FEEDBACK

Unlike feedback inhibition, the cell may stop producing enzymes in response to an excess intracellular concentration of the end product. This is often achieved by blocking the transcription of the genes encoding the biosynthetic enzymes. Genetic repression is the term used to describe this process. A repressor protein is activated by an amino acid that the cell has an excess of, and this protein prevents the biosynthetic genes from being further transcribed. Substrates, like carbohydrates, on the other hand, can activate the transcription of genes. We refer to this genetic process as induction. To control a metabolic route, many species may use varied combinations of feedback inhibition, repression, and activation(Sheen, 1994). The generation of energy is one of the goals of the metabolism of carbohydrates. Adenosine triphosphate (ATP) is the universal energy transfer substance present in living cells. The cell can produce ATP in two ways: (a) through substrate-level phosphorylation, which occurs during the breakdown of carbohydrates and involves the transfer of high-energy phosphate from a chemical compound (Phosphoenol pyruvate) to adenosine diphosphate (ADP); or (b) through oxidative phosphorylation, which uses the energy from an electrical and chemical gradient that forms across the cell membrane to drive an ATP hydrolase complex that is bound to the membrane, producing ATP from ADP and inorganic phosphate. A complex series of reactions is necessary to generate an electrical and chemical gradient (proton motive force) across the cell membrane. These reactions involve the transfer of H^+ and e^- from chemical intermediates of the Embden-Meyerhof and TCA cycles to a group of membrane-associated proteins known as cytochromes(Taylor & Kelly, 2019). The energy released during the transfer of e^- from one cytochrome chain member to another is utilized to expel H^+ from the cell. The resultant differential in pH (chemical potential) and charge (electrical potential) between the interior and outside of the cell can be used by the cell to produce ATP. A terminal electron acceptor molecule is necessary for the cytochrome system to function. Under anaerobic conditions, *E. coli* has alternative electron acceptors from which it can select based on availability (nitrate). Under aerobic conditions, oxygen will fulfill that function.

The oxidation of a sugar molecule throughout the process of hydrogen transfer from intermediates to molecules that may accept hydrogen is known as carbohydrate metabolism. Nicotinamide adenine dinucleotide (NAD) is often used as a hydrogen acceptor. The reduced form of NAD, or NADH, is what gives the cytochrome system access to H^+ and e^-. When a cell is forced to produce in an anaerobic environment without any alternative electron acceptors, all of the NAD may be transformed to NADH and NAD^+ could be completely depleted(Yang & Sauve, 2016). Because NADH, which is created in the early stages of glycolysis, is unable to transfer H, the cell is unable to manufacture new NAD^+. The cell would stop growing if this condition worsens since there may not be any NAD^+ available to carry out glycolysis. Many microbes, including *E. coli*, can replenish NAD^+ to survive. They do this by permitting NADH to transfer H to pyruvate or acetyl CoA, two final intermediates in the glycolytic process. Depending on the organism, the fermentation process yields lactic acid, isopropanol, butanol, ethanol, and other compounds. Not every fermentation process is carried out by *E.coli*. It is restricted to the formation of lactate, acetate, formate, ethanol, CO_2, and H_2. Glycolysis is not the only way that the cell breaks down glucose. Alternative metabolic pathways exist for the breakdown of glucose.

1.6 NITROGEN AND OTHER METABOLITES

Nitrogen is a structural component of every amino acid, purine, pyrimidine, and other compounds found in cells. Certain microbes, including Rhizobium, may fix atmospheric nitrogen by using nitrogenase to create ammonia (NH^{4+}), which they can subsequently absorb into amino acids(Bhatla & Lal, 2023). Other organisms need to begin with NH^{4+}, like *E. coli*. One of the 13 essential metabolites, α-ketoglutarate, is subjected to amidation during the absorption of nitrogen to produce glutamic acid. Transamination processes transfer the amino nitrogen to other molecules once it is assimilated into glutamate. Aspartate, for instance, is created when glutamate transfers its amino group to oxaloacetate. Aspartate, like glutamate, is a precursor for several other amino acids. Some bacteria, including Clostridium and Bacillus, create specialized structures known as endospores. Endospores offer protection against radiation, heat, desiccation, and other environmental elements that could endanger the organism's survival. Endospores offer a selection advantage for the survival and spread of the species(Paul et al., 2019). Spores exhibit a well-defined, multilayered exosporium (electron microscope), along with a dense outer coat that appears as a considerably darker region and a thick inner coat. The nuclear material and the darkly stained ribosomes may also be seen within the spore. The generation time, or the amount of time needed for a single cell to divide, has up until now been thought to be constant for every cell in the population. The cells first need time to acclimate to their new surroundings, and there is a delay in the time it takes for every cell to divide. There may be a decrease in the number of viable cells and some of the original inoculum cells may not make it through this lag period. Eventually, the surviving cells start to divide more quickly as they get used to their new surroundings. Until the conditions start to deteriorate, this rate will not change. This stage of growth is known as the logarithmic (log) phase, or more accurately, the exponential phase, because displaying the cell number logarithm during this time produces a linear curve.

In the stationary phase, every microbe culture finally reaches its maximum population density. The elements that lead to a decrease in growth rate are exhaustion of vital nutrients, buildup of harmful waste products, oxygen deprivation, or the establishment of an unfavorable pH. The number of cells that can divide (viable cells) is roughly equal to the number of cells that cannot divide (nonviable cells), even if cell division continues during the stationary phase. As a result, the balance between the number of cells that can divide and the number that cannot is represented by the stationary phase. The death phase starts as a result of the organisms in the population dying off and the viable population declining.

The characteristics of the organism being observed and the various variables that lead to cell death will determine the precise form of the curve during the death phase. When heat-induced death occurs, for example, viable cell counts decrease logarithmically, and the death phase may take on a linear shape. Bacteria are typically grown in batch cultures, which include inoculating a flask with medium and allowing growth to proceed. It is difficult to regulate the growth rate in this closed system. Bacteria can be grown at varying speeds in a batch culture as long as the added nutrients are provided at a concentration that does not promote maximum growth.

A chemostat is used to cultivate bacteria at high cell density and moderate growth rates(Ferenci, 2007). With this device, a reservoir holds a fresh medium that contains a limiting nutrient, which is fed to the culture vessel at a predetermined rate. An overflow mechanism that drains medium and cells at the same pace as new media is introduced maintains the culture vessel volume constant. The growth rate in a chemostat is set externally by adjusting the rate-limiting nutrient that is introduced to the culture vessel. The growth rate increases with the rate at which the limiting nutrient is provided.

1.7 BACTERIA

Lithotrophs obtain their energy from the oxidation of inorganic substances (chemolithotrophs) or light (photolithotrophs), using carbon dioxide as their only source of carbon. Organic molecules are synthesized using inorganic nitrogen. When it comes to sources of carbon and energy, organotrophs typically favor organic substrates. As a source of energy, light is used by photoorganotrophs to assimilate organic molecules and carbon dioxide. Chemoorganotrophs grow by using chemical substances. Despite minimal dietary needs, chemolithotrophic bacteria need to have complicated metabolisms since they synthesize all of their cellular components and get their energy from the oxidation of inorganic molecules. The inability of chemolithotrophs to develop or digest foreign organic substances is one of their basic traits. Exogenous organic carbon sources can be utilized by facultative chemolithotrophs. Chemolithotrophs have special carbon dioxide fixation systems, such as the reductive carboxylic acid (Campbell-Evans) cycle and the ribulose bisphosphate (Calvin-Benson) cycle. While some organotrophic organisms can obtain their carbon from carbon dioxide, most species prefer to obtain their carbon from organic sources and are generally unable to survive on carbon dioxide alone.

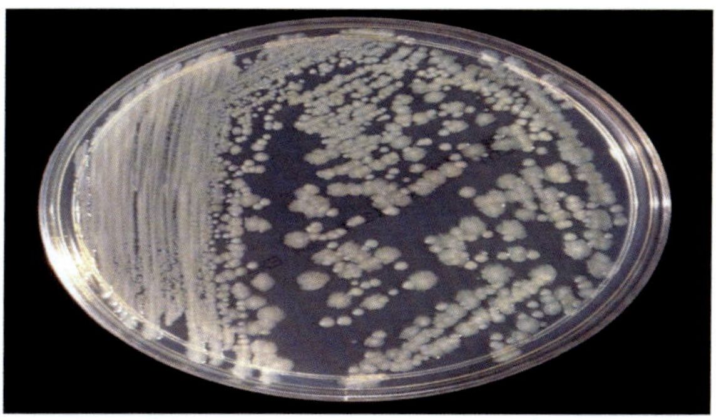

Figure 1.7 *Enterobacter aerogenes*

Source: https://upload.wikimedia.org/wikipedia/commons/7/79/Enterobacter_cloacae_01.png,*Creative Commons License.*

Although many members of the group grow better when provided with organic nitrogen molecules, organotrophs can consume inorganic nitrogen as well. For instance, when glucose is the only organic nutrient available, E. Coli, *Enterobacter aerogenes*(Figure 1.7), yeasts, and molds survive. Certain nitrogen sources, such as amino acids, purines, or pyrimidines, are also required by other organotrophs, including staphylococci and streptococci. Certain organisms require fatty acids, especially when certain B vitamins are not present. Many growth factors, including amino acids, purines, pyrimidines, and other organic elements, have been shown to have this kind of activity. A nutrient cannot be considered a growth requirement if it can be manufactured in the presence of the necessary vitamin, even if it can entirely replace an organic nutrient in a specified medium.

Sterols are necessary for the membranes of certain Mycoplasmataceae species, although they are absent from the majority of bacterial membranes(Razin, 1981). Sterols are necessary for Mycoplasma development. Although Acholeplasma do not require sterols, they do create terpenoid molecules that have similar functions to sterols(Ourisson et al., 1987). Although sterols are present in the cell membranes of fungi (yeasts and molds), they also seem to be able to synthesize them in most situations. Aerobes are microorganisms that need oxygen to carry out their energy-producing metabolic processes; anaerobes are those that are unable to do so. Depending on the amount of available oxygen in their culture, facultative organisms can use fermentation or respiratory activities. Cytochrome oxidase and cytochromes are found in aerobic organisms and are involved in the process of oxidative phosphorylation. Water is one of the byproducts of respiration, and oxygen (O_2) acts as the terminal electron acceptor. Certain oxidation-reduction enzymes react with molecular oxygen to produce hazardous compounds such as hydrogen peroxide (H_2O_2), hydroxyl radicals ($OH^{\cdot}$), and superoxide ($^{\cdot}O^{2-}$)(Kordrostami et al., 2019). Catalase is typically not produced by aerotolerant species. Therefore, culturing on a medium containing blood or other organic materials that possess catalase or peroxidase activity is often used to increase the growth of these organisms. Because they do not interact with molecular oxygen or produce peroxide or superoxide, oxygen-free organisms may be able to withstand it. Many natural environments, including dirt, marshes, and deep lakes, as well as the regular microflora of animals and humans, contain anaerobic bacteria belonging to different genera.

Analyzing a microorganism's physiology used to be a laborious procedure. We still lack a comprehensive understanding of the biochemistry and genetics of *Escherichia coli*, the common intestinal microbe, even 60 years after the analysis of this organism began.

Mycobacterium leprae (Figure 1.8) and *Rickettsia prowazekii* (Figure 1.9), both are incapable of growing outside of a living host (Schmutzhard & Helbok, 2014). We can now confidently identify the biochemical pathways they have, their energy production mechanism, potential virulence proteins, and their evolutionary relationship to other bacteria.

All of this is the result of advanced sequencing methods that have made it possible to decode entire genomes in comparatively short amounts of time. Strong sequence analysis software has been developed that can identify promoter sequences and open reading frames (ORFs) and compare newly discovered sequences with those that are already known. A homology search, which compares the amino acid sequence of an ORF with an unknown function to all other known protein sequences, finds the proteins that have motifs and sequences comparable to those found in the ORF. DNA sequencing can start as soon as a random clone library is created. Each single strand (clone) is utilized as a template to create a sequence of fragments, each of which is only one base longer than the one before it. Using a short oligonucleotide primer (20–25 bases) that binds to a sequence in the vector molecule is the first step. Keep in mind that DNA-dependent DNA polymerases can only assemble DNA by appending a new nucleotide to a developing strand 3' hydroxyl group. Each primer molecule permits DNA sequencing to start at a fixed place.

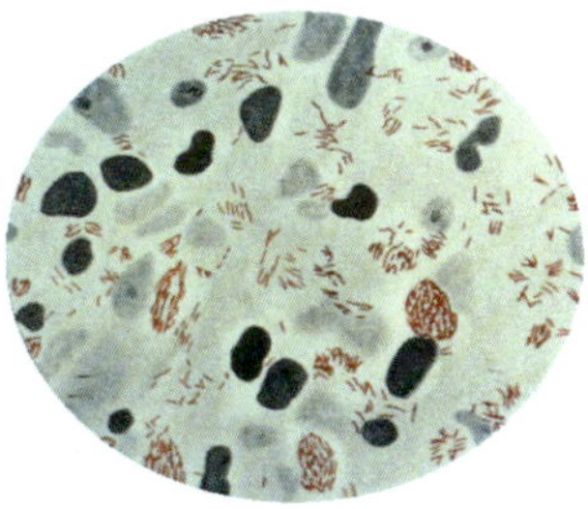

Figure 1.8 *Mycobacterium leprae*

Source: https://upload.wikimedia.org/wikipedia/commons/d/da/Mycobacterium_leprae.jpeg,*Creative Commons License.*

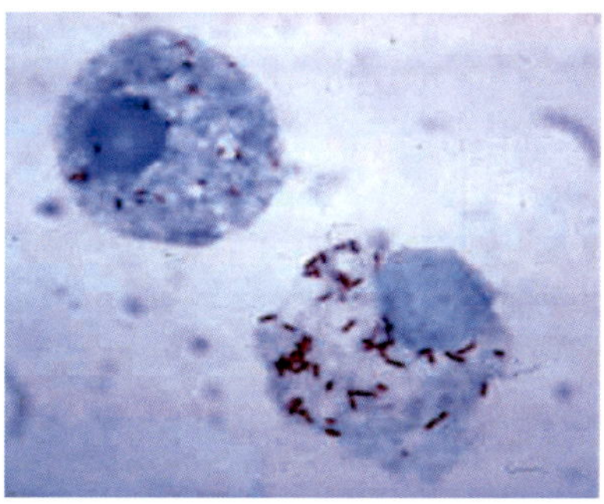

Figure 1.9 *Rickettsia prowazekii*

Source: https://upload.wikimedia.org/wikipedia/commons/8/86/Rickettsia_rickettsii.jpg,*Creative Commons License.*

1.8 DIAGNOSTIC TOOLS

Minute amounts of fluorescently tagged, dideoxy versions of each nucleotide are added to the DNA reaction mix, along with regular deoxynucleotides, to generate fragments of varying sizes while precisely labeling the final nucleotide integrated into each fragment. The combination of diverse-sized and tagged DNA fragments is separated by size using polyacrylamide gel electrophoresis to read the sequence. All DNA fragments move toward the positive node due to the total negative charge on the molecule, and smaller DNA fragments move more quickly than larger pieces due to the polyacrylamide matrix reaction. As a result, there is a movement of fragments that gradually move down the gel, each exactly one base longer than the last. The fluorescent tag at the end of each fragment will be excited/activated by an optical laser pointed toward the gel end, and the color released will be interpreted as a certain base. Shotgun cloning creates fragments with overlapping ends, which can be identified and assembled into contigs by computer analysis of the many DNA sequences(Summer, 2009). The contigs are put together into a full genome that is ready for annotation with a minute gap sealing. It is believed that processes like gene duplications (within an organism), horizontal gene transfers (HGT) between species, and mutational divergence of genes have significantly influenced evolution. These processes have also contributed to the rapid diversification of reactions catalyzed by enzymes and have provided building blocks for the development of new enzyme activities. Orthologous genes (or proteins) are those that have evolved from the same ancestral gene in two distinct animals (species). They might not still have an ancestral purpose. Gene duplications are reflected in the emergence of paralogous genes (or proteins). For instance, the gene that codes for σ-38 and the gene that codes for σ-70 are paralogs. Duplication of one gene gave rise to another, which then underwent mutational divergence to modify something (like promoter selectivity).

Gene array analysis can be used to directly compare the gene expression patterns produced by a single strain growing under two different environments. This is accomplished during the synthesis of cDNA by adding several fluorescent-tagged dTTP solutions into the reaction mixtures. This allows the cDNA generated from cells grown under various conditions to be labeled with distinct fluorescent markers. Next, a combination of the two differently labeled cDNA preparations is probed using a single array. A location on the array will flash one color (red) if a certain gene is expressed under only one condition. This is because only the RNA extracted from cells grown in that condition will include that species of RNA. A second gene location on the array will display the opposite color (green) if it is only produced when cells are growing under the other condition. Spots corresponding to genes that are expressed equally under both conditions will display a third color (yellow/orange), which is a combination of the first two colors when a computer-generated composite image of the two scans is created. All of the expressed proteins are included in its proteome. The proteome is dynamic, whereas the genome remains constant due to variations in the extracytoplasmic or cytoplasmic environments(Antelmann et al., 2005).

When a cell experiences a stressor like heat shock or acid shock, this is most obvious. An extensive rearrangement of the proteome occurs in reaction to an approaching threat. In reaction to the stress, many new proteins are expressed while the amounts of existing proteins decrease. Although transcriptional fluctuations can cause changes in the proteome, translational or posttranslational forces cannot be predicted by gene array technology. This means that although a change in growth conditions may not alter a gene's transcription, it may have a significant impact on the translation of its mRNA. Two-dimensional polyacrylamide gel electrophoresis is commonly used to view the proteome. Isoelectric focusing (IEF), a technique that divides proteins according to their net charges, is used initially for a cell extract that has a blend of 1000–2000 distinct proteins. In IEF, the current is applied to an immobilized pH gradient gel containing proteins. Many side groups of amino acids, such as carboxyl or amino groups, are present in every protein and might display a charge based on whether or not they are protonated. The pKa of the side groups and the pH of the surrounding environment determine how a side group alters its protonation state (charge). Side groups will often undergo protonation at pH levels lower than their pKa value. The amino group of lysine will be positively charged (protonated) at pH 7, but the carboxyl group of glutamate will be negatively charged (unprotonated). The total of the side-group charges determines a protein net charge. Due to their initial charges, the proteins will start to migrate toward the opposing poles of the IEF gel; however, throughout this migration, each protein is exposed to a pH gradient. The pH encountered at a particular spot in the gel eventually causes an equal number of positive and negative charges on the protein, resulting in a net charge of zero (isoelectric point). The protein stops migrating and concentrates at that spot since there is no charge. The first dimension in two-dimensional gel electrophoresis is called IEF separation. Even though the first stage of separation is effective, proteins with different molecular weights do have the same isoelectric points. As a result, ten or fifteen distinct proteins may be present in one band on an IEF gel. Proteins must be separated according to their molecular weight using a second method to provide a more comprehensive image of the proteome. The strip from the first dimension is first soaked in a sodium dodecyl sulfate (SDS) solution to initiate the second dimension. All proteins will be coated with the negatively charged molecule SDS, which will also give them a negative charge.

After that, the strip is positioned above a vertical slab gel with a polyacrylamide matrix and is once more exposed to an electric current. The proteins start to go toward the positive pole through gaps/spaces in the matrix. Larger proteins move more slowly across the gaps, while smaller proteins, those with a lower molecular weight, move more easily. The proteins are separated by molecular weight due to this sieve action.

1.9 PROTEINS

The proteins can be observed by staining them with Coomassie blue dyes or fluorescent-tagged dyes, or by radioactively labeling them with 35S-methionine(Wirth & Romano, 1995). Proteomes generated in response to various growth conditions can be compared using computer analysis. A protein needs to be identified before it may be studied further. The protein of interest can be removed from the gel and its specific mass can be ascertained using mass spectrometry when the genome of the organism or a closely related organism has already been sequenced. This method involves breaking down a protein using a protease like trypsin and then introducing the fragments into a mass spectrometer to determine the exact mass of each fragment. When the entire genome has been sequenced and the precise molecular weights of all putative peptides have been determined, this can be utilized to definitively identify the protein. The spot can be removed and subjected to N-terminal amino acid sequence analysis or cleaved using a protease, like trypsin, and the fragments separated by HPLC and sequenced, if the genome has not yet been sequenced. The resulting peptide sequence or sequences are then utilized to query other databases to find matches. As an alternative, partial sequence information for a peptide can be obtained using mass spectrometry, or MS/MS. After additional fragmentation, the locations of certain amino acids in the peptide are determined using the mass spectrum obtained from the overlapping fragments. One can utilize the generated amino acid sequence to search for matches in databases that are known. After a mutant is discovered, the encoding gene can be located, cloned, sequenced, and the biochemical step in the process investigated. Thousands of mutants can be screened on a single plate if there is a positive selection phenotype, for resistance to an amino acid analog or a fermentable (red colony) versus nonfermentable (white colony) phenotype on a MacConkey medium.

An effective method for examining different facets of gene expression is joining readily analyzed reporter genes, like β-galactosidase or green fluorescent protein, to target gene promoters. When a promoterless lacZ reporter gene is inserted into a target gene in a way that puts lacZ under the control of the target gene promoter, the result is known as an operon or transcriptional fusion. Therefore, the variables that govern the expression from the promoter of the target gene will likewise govern the generation of β-galactosidase.

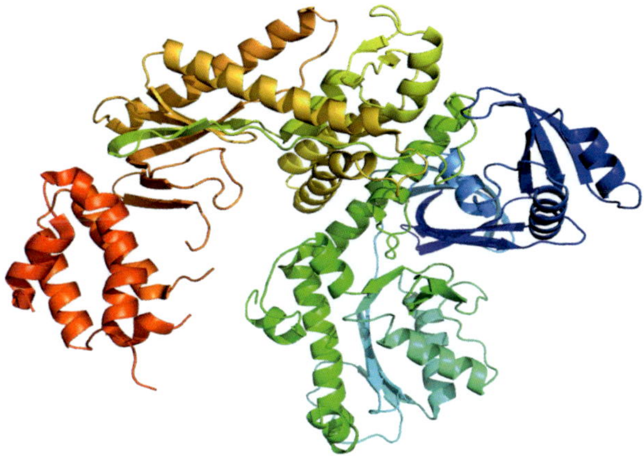

Figure 1.10 DNA polymerase

Source: https://upload.wikimedia.org/wikipedia/commons/a/a2/Pol2_structure_%28Based_on_35KM%29.png,*Creative Commons License.*

The ribosome-binding site of the lacZ message that results from this fusion is still present, therefore transcriptional control is usually responsible for the regulation. The gene or protein fusion, is introducing a reporter, such as lacZ, into a target gene that lacks a ribosome-binding site and its promoter. Anything regulating the target gene transcription or translation will also regulate the amounts of β-galactosidase. Sometimes, to construct these fusions, a reporter lacZ gene is randomly inserted into the Salmonella or *E. coli* genome using a genetically modified μ phage(Chen et al., 2021). As an alternative, the fusion can be created in vitro and then introduced into recipient cells in a way that encourages allelic exchange of the resident gene for reporter fusion. Furthermore, these fusions produce useful phenotypes that can be used in mutant searches to find novel regulators.

A thermostable DNA polymerase (Figure 1.10) uses primers to copy the target DNA. The thermophilic bacterium *Thermus aquaticus* produces the enzyme taq polymerase, which is widely used in polymerase chain reactions (PCRs). The reaction mixture needs to be repeatedly heated to 95 °C (to separate DNA strands), cooled to 55 °C (primer reannealing), and heated to 72 °C (ideal reaction temperature) to amplify a specific-size DNA. This requires the use of thermostable DNA polymerases. Since 25 to 30 heating and cooling cycles are needed for amplification, a thermocycler is a device that can swiftly and reliably provide these cycles. Many modifications have been made to the PCR process for various uses. PCR can be used to create site-directed mutagenesis-specific mutations in a certain gene. For instance, a possible functional motif (Zn^{2+} finger) can be discovered through computer analysis of a gene.

It is crucial to determine whether a putative transcriptional regulatory protein directly binds to the target DNA sequence when a gene encoding the protein has been found. One method is to mix a pure protein with the suspected target DNA fragment and observe if the protein alters the fragment's electrophoretic mobility. Molecules with large sizes move more slowly than those with less. When a protein attaches itself to a DNA molecule, the resulting complex is larger than the DNA and moves across the gel more slowly. This is referred to as a gel shift or mobility shift. It can be used to ascertain whether a suspected DNA-binding protein is indeed a component of a protein combination that induces a gel shift. When an antibody against the putative DNA-binding protein is added to the binding reaction, the target segment will undergo a supershift. A larger gel shift (supershift) results from the complex of anti-DNA-binding protein, DNA-binding protein, and DNA than from the binding protein–DNA complex. The putative protein was not the cause of the first shift if there was no supershift. One technique to accomplish this is primer extension. A reverse transcription procedure is performed to extend a DNA primer to the 5' end of the message by annealing it to the mRNA transcript. Because radioactive nucleotides are included in the reaction mixture, this results in the production of radiolabeled cDNA. The use of the same primer in a DNA sequencing reaction with the template DNA is crucial to this process. On a polyacrylamide gel, the primer extension fragment is subsequently run next to a DNA sequencing ladder. The primer extension fragment dimensions will match those of one of the sequencing ladder rungs. The transcriptional initial site is correlated with the nucleotide that rung represents in the sequence.

1.10 GENE EXPRESSION

Techniques for identifying DNA, RNA, and protein are vital tools for examining gene expression patterns(Lockhart & Winzeler, 2000). A Southern blot starts by cutting a DNA molecule with one or more restriction endonucleases. Following agarose electrophoresis, the molecules are moved from the agarose gel (AG) to a nitrocellulose membrane (NM) that is placed on top of the gel. Capillary action is used in the transfer procedure to draw a buffer through the agarose gel and subsequently the facing membrane. The buffer-transported DNA is deposited on the membrane in precisely the same pattern as it appears in the gel. After the DNA fragments are adhered to the membrane, the strands are chemically separated, and the membrane is probed using a tagged, denatured

DNA fragment under hybridizing conditions that let the matching DNA fragment anneal with the probe. A fluorescent dye, radioactivity, or biotin can be used to label the probe. Afterward, the biotin is found using a chemiluminescent enzyme assay. Autoradiography is the process of exposing the membrane to X-ray film and photographically processing the film to visualize DNA fragment bands that have annealed to the probe. Northern blots are used to examine whether a particular RNA molecule is present, how large it is, and how it is processed in a cell extract. Special care is to be taken when extracting RNA from cells to prevent the preparation from becoming contaminated by RNases, which are commonly found in laboratory settings. As with Southern gels, the RNA is separated by electrophoresis on an agarose-formaldehyde gel, and the fragments are then deposited onto a nylon or nitrocellulose membrane. A small radiolabeled DNA probe tailored to a specific mRNA can then be used to probe the membrane. Following hybridization, the membrane undergoes autoradiography or phosphorimaging, a process that captures the energy released by gels or membranes as light or radioactivity.

The yeast two-hybrid system is the currently accepted approach. Using this method, genes encoding two possibly interacting proteins are fused to the individual yeast GAL4 transcriptional activator components. These are the transcriptional activation domain fragment and the DNA-binding domain fragment. While different reporter genes can be employed, the interaction between the two parts of the GAL4 activator is required to initiate the expression of a yeast chromosomal GALI-lacZ fusion(Griggs & Johnston, 1993). Further, β-galactosidase, whose enzyme activity is easily tested, is encoded by the lacZ gene. Normally, a single GAL4 protein has both domains. Nevertheless, if each domain is fused to one of a pair of additional interacting proteins, the two separated domains can still interact. The interaction between the two target protein domains of the resulting hybrid proteins brings the GAL4 DNA-binding and activation domains together. After being put together, the two-hybrid GAL4 complex will connect to GALI-lacZ and start its transcription. Next, GALi-lacZ expression can be observed on agar plates with X-gal, a chromophoric β-galactosidase substrate, or it can be examined in vitro directly. This method's strength lies in its ability to recognize unknown proteins that interact with known proteins.

Regulatory regions are the sequences that are next to the structural gene or actual coding region. These areas consist of an operator region, which is where a regulatory protein binds, and the promoter, which is where transcription starts. Positive control is achieved by regulatory proteins, whilst negative control is achieved by blocking transcription. While activator proteins are linked to positive control, repressor proteins generate negative control. Three stages are necessary for transcription initiation: RNA polymerase binding, isomerization of a few nucleotides, and RNA polymerase leaving from the promoter region to permit message elongation. Negative regulators often prevent binding, although activators interact with RNA polymerase to increase the likelihood of one or more steps. An operon is a group of several different genes arranged in tandem and regulated by a single regulatory region (lac operon). An operon produces a polycistronic message since it contains all of the structural gene information on a single mRNA molecule. Since all of the transcriptions of these genes depend on a shared regulatory region, the transcriptions of each operon either increase or decrease concurrently, resulting in coordinated regulation of these genes. Genes involved in specific metabolic pathways are frequently dispersed throughout the bacterial chromosome rather than in an operon. However, because they all react to the same regulatory protein (arginine regulator), they may be regulated. Regulons are systems that contain dispersed genes that are regulated.

Chapter 2

PHYSIOLOGY OF BACTERIA

Contents

2.1 INTRODUCTION

In eukaryotes, the term cytoskeleton has evolved to refer to a variety of filamentous cell structures, some of which are dynamic and others of which are more stable. The majority of these filament types are made up of a single protein component that can assemble into polymers both in vivo and in vitro. These filament systems play a major role in cellular organization by regulating cell division, the transport of molecules, vesicles, and organelles, as well as the maintenance of cell shape and mechanical strength. A protein that belongs to one of the three cytoskeletal protein superfamilies, tubulin, actin, or intermediate filaments (IFs), which include keratins, lamins, and other specialized proteins that form more static filaments, is what makes up a cytoskeletal filament in eukaryotes. Tubulin and actin form dynamic actin filaments and microtubules, respectively. Because of their small cells, bacteria have long presented challenges to cell biologists researching cellular organization. Because of their microscopic size, which exceeds the resolution of optical microscopes, bacteria were long believed to be devoid of cytoskeletal components. There were only a few known interior structures in bacterial cells, and they appeared to be unclear and peculiar to specific organisms.

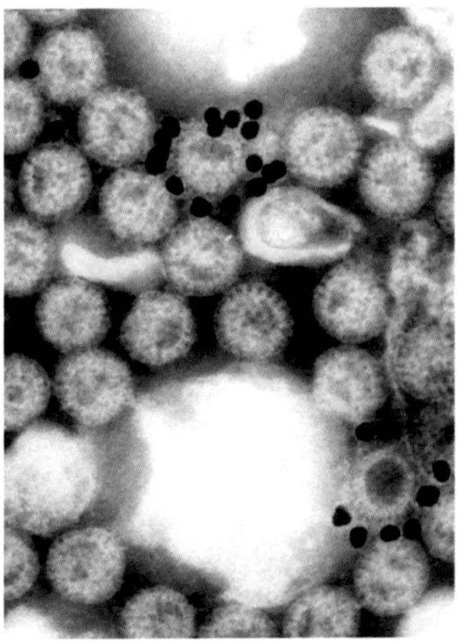

Figure 2.1 Immunoelectron microscopy

Source: https://upload.wikimedia.org/wikipedia/commons/0/0f/Rotavirus_with_gold-_labelled_monoclonal_antibody.jpg,*Creative Commons License.*

2.2 TECHNIQUES FOR STUDYING

Bi and Lutkenhaus used immunoelectron microscopy (Figure 2.1) to find that, in *Escherichia coli*, a well-conserved protein known as FtsZ associated with cell division localized in a distinctive pattern at the center before any discernible septum invagination (Addinall & Lutkenhaus, 1996). Moreover, FtsZ maintained a pattern compatible with a constricting ring structure at the leading edge of the enclosing septum during cell division. Since then, many bacteria have been found to have FtsZ rings, which are assumed to provide the fundamental cytoskeletal framework supporting the division apparatus. Technological developments in optical microscopy, such as the utilization of time-lapse imaging, deconvolution analysis, and green fluorescent protein (GFP) fusion technologies, have increased our understanding of the organization of bacterial cells. According to recent developments in cryo-electron tomography techniques, even more information can be observed. Eubacteria that generate filamentous structures with cytoskeletal roles have three of the known eukaryotic cytoskeletal proteins: tubulin, actin, and IFs. It is important to understand the biophysical and biochemical properties of the prokaryotic cytoskeletal proteins, which include FtsZ, a tubulin homolog; two specialist tubulin homologs (BtubA and BtubB) found in Prosthecobacter species; a variety of bacterial actin-like proteins, including MreB, which aids in the maintenance of cell shape in many rod-shaped bacteria; some actin-like specialist elements, such as ParM and MamK filaments; and a single IF-like protein, represented by crescentin, which is involved in the defining cell shape in *Caulobacter crescentus*. Walker A cytoskeletal ATPase, or WACA proteins, is a novel family of cytoskeletal proteins that has been characterized in bacteria. The eukaryotic counterparts of this class are still unknown. This family of proteins includes ParA, MinD, Soj, and MipZ.

Many tubulin proteins, such as α-, β-, γ-, δ-, and ε-tubulin, are expressed by eukaryotic cells(Oakley, 2000). The two primary tubulins that makeup microtubules are α- and β-tubulin. All eukaryotes have microtubules, which are vital cytoskeletal components needed for cell division and a variety of intracellular transport processes. During mitosis, microtubules create the mitotic spindle, which serves as the framework for the separation of daughter chromatids toward opposing cell poles. Typically, microtubules consist of thirteen filaments. Each filament is composed of αβ-tubulin heterodimers that are connected longitudinally and laterally into a tube-like structure. Two functional domains, comprising two β-sheets encircled by α-helices, constitute the core structures of α- and β-tubulin. Both domains' N-terminals have a GTP binding site and a Rossmann fold that is akin to many ATPases. Some catalytic residues for GTP hydrolysis are present in the C-terminal domain, which shares structural homology with the family of proteins similar to chorismate mutase. The tubulin proteins are GTPases, and the essential amino acid residues from both subunits are used to construct the active site at the interface between subunits. The nucleotide-binding site of one subunit is provided by its

N-terminal domain, while the T7 loop, which contains the residues required for nucleotide hydrolysis, is provided by the C-terminal domain of another subunit. As a result, GTPase activity is limited to situations in which two or more subunits connect. The α and β subunits of tubulin alternate, with GTP binding sites between each, to form microtubules. These heterodimers are connected to form tubulin αβ heterodimers. With β-tubulin always present at one end (plus end) and α-tubulin at the other end (minus end), this arrangement produces a distinct polarity. Dynamic instability in microtubules causes the filaments to rapidly expand or contract. Three biological characteristics give rise to this trait. Firstly, the filaments can only assemble and disassemble from their ends; subunit exchange does not take place within the filament. Second, nucleotide exchange inside the filament is restricted due to the sealed nucleotide-binding pocket between tubulin monomers. Third, a destabilizing conformational shift inside the filament has been suggested to be induced by GTP hydrolysis, leading to a bent shape. The GDP bent form of the filament quickly disintegrates. Nonetheless, a GTP cap at the end of the microtubule can stabilize the filament, keeping it in the straight conformation and allowing it to extend. If the GTP-cap hydrolyzes to GDP, the filaments break down naturally. Therefore, whether or not microtubules grow depends critically on the condition of this cap.

2.3 CELL DIVISION PROTEIN STUDIES

The localization of FtsZ into a ring structure (Z ring) is the first known step in cell division in bacteria(Weiss, 2004). Several proteins, including FtsZ, are involved in cell division. By isolating conditional mutants that exhibit filamentous cellular morphologies when cultivated at a temperature that is not permissible, many of these cell division proteins were found. As a result, many proteins involved in cell division have been referred to as filamentous temperature-sensitive proteins, or fts. Following Z ring formation, many additional membrane-bound, cytoplasmic proteins that are involved in cell division are localized. The device that divides cells is known as the divisome or septasome, and it is the culmination of all these proteins assembled around the Z ring. Understanding the function of FtsZ has been at the forefront of most of the research on the molecular mechanisms involved in bacterial cell division and the involvement of the divisome. The division process depends on FtsZ, the first protein that has been shown to localize to the division site. Furthermore, the majority of bacteria and archaea exhibit remarkably high levels of conservation for FtsZ, indicating that it serves a key biological function. Only the following are believed to lack FtsZ: *Ureaplasma urealyticum*, (Figure 2.2) a species of Pirellula, the Crenarchaeota, and the Chlamydiaceae family. FtsZ is also present in these organelles in some eukaryotes, supporting the endosymbiont model for the origins of chloroplasts and mitochondria. Early in the 1990s, researchers discovered a conserved tubulin pattern within the FtsZ sequence(Addinall & Holland, 2002). This finding gave rise to the theory that FtsZ and tubulin had an evolutionary link. This was unexpected given that amino acid sequence alignments reveal only an average of 10–18% sequence identity between the proteins. The controversy was resolved in 1998, though, when the three-dimensional protein structure of FtsZ was revealed. It was discovered that FtsZ shared the same two-domain fold as tubulin, with the residues responsible for GTP hydrolysis placed on a flexible loop in the C-terminal domain known as the T7 loop, and a conserved GTPase binding pocket in the N-terminal domain.

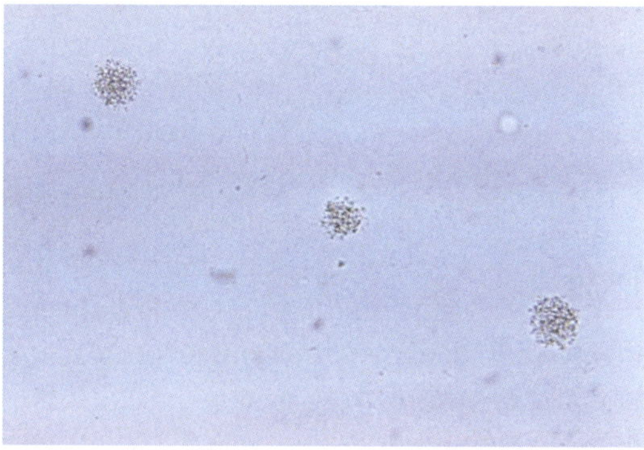

Figure 2.2 *Ureaplasma urealyticum*

Source: https://upload.wikimedia.org/wikipedia/commons/d/d6/10931_lo-res_%281%29.jpg,*Creative Commons License.*

FtsZ assembles into a highly organized structure in vivo that is consistent with its connection to tubulin, albeit its exact molecular architecture is unknown. Cryo-electron tomography has shown structures that may be FtsZ filaments, while further data is needed to validate this interpretation. Fluorescence recovery after photobleaching or FRAP experiments has shown that the Z-ring structure is highly dynamic and is constantly remodeled or rebuilt in vivo, with FtsZ subunits exchanging between the ring and the cytoplasm in a matter of seconds. Time-lapse live-cell microscopic imaging of FtsZ–GFP fusions in both wild-type and mutant bacterial strains has also revealed many reports of moving ring and helical structures assembled from FtsZ in vivo. The Z ring dynamic behavior indicates that it is a versatile and flexible structure. FtsZ self-assembles into linear chains of molecules that associate, known as protofilaments, in vitro. Protofilaments have been observed in a variety of morphologies, including straight and curved forms. Moreover, FtsZ protofilaments unite to form sheets, tubules, and bundles, among other higher-order structures. These assemblies have been reported to form in a variety of situations and nearly every known nucleotide-bound state, such as GTP, GDP, and the non-hydrolyzable guanylyl-(α,β)-methylene diphosphate. It is still unclear how GTP hydrolysis affects FtsZ function, but in vivo mutants of the protein can support cell division even when their GTPase activity is reduced.

2.4 FtsZ PROTEIN AND ASSEMBLY DETAILS

The *in vitro* polymer form of FtsZ that most closely resembles the native form is unknown, and the situation is further complicated by the many proteins that are known to interact with FtsZ during divisome assembly, potentially changing the biochemical properties of FtsZ. Insights into the probable configuration of individual FtsZ subunits within the polymer have been gained from structural characterization at the molecular level of protofilaments generated by FtsZ *in vitro*. The head-to-tail assembly of FtsZ monomers within protofilaments is comparable to that of αβ-tubulin within microtubules. Every FtsZ monomer in a protofilament maintains longitudinal connections with organized FtsZ molecules (10° tilt) above and below, ensuring that every molecule is aligned in the same direction. Therefore, longitudinal bonds between FtsZ molecules are the primary interactions that generate single protofilaments. A catalytic loop for GTP hydrolysis, contributed by a neighboring FtsZ molecule in the protofilament, and a nucleotide-binding pocket, contributed by one FtsZ molecule, are inserted between two FtsZ molecules. Some conserved aspartate residues and a glutamine residue that coordinates a magnesium ion are found in the catalytic loop. Nucleotide hydrolysis cannot occur unless FtsZ self-assembles. Significant distinctions exist between the nucleotide-binding pockets of tubulin and FtsZ. Because nucleotide exchange is restricted and tubulin nucleotide-binding pockets are sealed in its microtubule-assembled form, tubulin assembly is dynamically unstable. On the other hand, FtsZ nucleotide-binding pockets facilitate nucleotide exchange more easily. This property may have a significant impact on the dynamic behavior of FtsZ filaments. FtsZ filaments continuously rearrange in vitro. *In vitro* observations of the joining of FtsZ filaments and depolymerization of FtsZ from within filaments further imply that nucleotide exchange takes place at internal locations, at least in a single protofilament(Wagstaff et al., 2017). It is generally believed that the Z ring is made up of many filaments, and the concentration of FtsZ in the cell is high. The exact composition of the Z ring has not been determined experimentally. Based on information gathered from the FRAP studies, it was calculated that roughly 30% of the cellular FtsZ appears to be present in the Z-ring structure at any one moment, which is consistent with the multifilament hypothesis for Z rings. Lateral interactions between neighboring protofilaments would likely be crucial for the formation and stability of the Z ring if it were made up of multistranded FtsZ filaments. In fact, in the case of tubulin, tubulin interactions with accessory proteins and the formation of microtubules depend on the lateral bonds between protofilaments. FtsZ lateral bonds are probably crucial for both connecting with the cell membrane and other proteins. Even though many proteins, including FtsA, ZapA, and MinC, are known to directly interact with FtsZ, less is known about the molecular structure of these interactions.

It is important to know that many ftsZ genes are encoded in several bacterial and archaeal genomes. This could offer some insights into the tubulin superfamily. Furthermore, plasmids of several Bacillus species have been shown to contain ftsZ genes recently. This is unexpected because these organisms live without these plasmids and have copies of the FtsZ chromosome on their chromosomes. This initiates the question of what function the FtsZ encoded in the plasmid serves. According to preliminary research, at least one of these plasmid genes is thought to be involved in the replication and/or segregation of the plasmid because the plasmid must survive in its host. Compared to FtsZ, such a function would be more akin to tubulin function. There are two classes of proteins involved in cell division: those that most likely directly contribute to the divisome function, such as constructing the division septum, and those that regulate the divisome, such as deciding when and where the Z ring assembles. The transmembrane proteins FtsK and FtsN are among these proteins. The cytoplasmic proteins FtsA, ZipA, ZapA, SepF, and EzrA are also included. In *E. coli*, FtsZ assembly is necessary for protein assembly in the mid-cell stage, which mostly takes place in steps. Membrane-bound proteins assemble after the cytoplasmic proteins. While FtsK in *E. coli* is used for chromosome segregation by facilitating the complete separation of chromosomes into the new cells created by the division septum, other assembling proteins, like FtsI/PBP2B, are known to be involved in cell wall (peptidoglycan) biosynthesis.

2.5 DIVISOME

To assemble at the proper time and location within the cell, the divisome needs signals(Aarsman et al., 2005). It is known that a variety of proteins interact with FtsZ in a way that is thought to control its location. Expression of SulA, an inhibitor of cell proliferation, is part of the SOS response of *E. coli* to DNA damage. To prevent Z-ring construction, SulA binds to the FtsZ polymerization interface and moves away from monomeric FtsZ. The Min proteins are the proteins that are known to impact Z-ring assembly. The Min proteins are a component of a system that prevents the divisome from assembling improperly close to the cell poles in rod-shaped bacteria. The MinC component directly inhibits Z-ring construction by interacting with FtsZ; however, the specifics of this interaction are yet unknown. Only cell division at the cell poles is inhibited by MinC subcellular localization, which is established by the other Min proteins. However, the precise methods by which EzrA, SlmA, ZapA, Noc, and MipZ interact with FtsZ to perform their stabilizing or destabilizing roles are unclear. It is believed that MipZ and ZapA stabilize the Z ring. On the other hand, it is believed that Noc, SlmA, and EzrA destabilize the Z ring. It is yet unknown whether these proteins function through protofilament bundling, FtsZ polymerization, or some other indirect method. Prokaryotes contain other tubulin-like proteins besides FtsZ. Two peculiar tubulin homologs named BtubA and BtubB are expressed by the bacterium genus Prosthecobacter(Yee et al., 2007). Since these proteins are unique to this species of bacteria, they most likely form a specialized cytoskeletal component. The crystal structures of BtubA and BtubB indicated that each protein has mixed properties of α- and β-tubulin and cannot be allocated to either α- or β-tubulin. However, both proteins demonstrate a closer link to eukaryotic tubulin than to FtsZ. BtubA and BtubB self-assemble in vitro and take on a filamentous shape like that of FtsZ and αβ-tubulin, albeit their exact functions are unknown. They may be the result of horizontal gene transfer events based on their divergence from eukaryotic tubulin.

2.6 ACTIN AND ITS ASSOCIATED PROTEINS

The sequence and function of the actin family of ATPases are varied. Members of this family are related by conserved sequence motifs involved in nucleotide binding as well as an ATPase structure. Since not every member of this family may polymerize or play a function in determining the shape of the cell, they are not all categorized as cytoskeletal proteins. Prokaryotes contain a variety of actin-like proteins, the majority of which were discovered via a sequence homology search using the catalytic core that hexokinase, actin, and the hsp70 proteins all share. Prokaryotes contain MreB, MreB-like proteins (Mbl and MreBH), ParM (StbA), FtsA, ActA, DnaK (hsp70), and the hexokinases as actin-like proteins(Graumann, 2001). It appears that MamK is a unique protein present exclusively in magnetotactic species. It is believed that MreB, ParM, and MamK play critical roles in plasmid segregation, cell shape determination, and the assembly of specialized cytoskeletal components. Nearly all eukaryotic cells contain the protein actin, which is prevalent. It creates a dynamic network that is used by different motor proteins to move molecules, vesicles, and organelles throughout the cell. Actin plays a major role in regulating the shape and facilitating the movement of certain types of cells. For instance, actin filaments play a direct role in the development of pseudopodia, which are essential to the process that permits amoeba movement. Actin and myosin together make up a vital component of the system that permits muscle cell contraction. Actin is composed of two domains, I and II, which further are subdivided into A and B subdomains. The two larger ones, identified as IA and IIA, consist of a β-sheet with five strands, encircled by three α-helices. The two smaller subdomains, IB and IIB, contribute some of the unique characteristics of each protein and vary in size and structure throughout the actin family. A highly conserved ATP-binding pocket between the two domains contains necessary aspartate residues that, when combined with either Mg^{2+} or Ca^{2+}, bind and hydrolyze ATP, which is crucial for the disintegration of F-actin filaments.

The actin subunits come together to produce F-actin, a dynamic helical polymer that resembles linear filaments. Each actin subunit experiences structural alterations during polymerization, which are partly aided by the rotation of domains I and II relative to one another. Actin has a characteristic asymmetry due to the head-to-tail arrangement of monomers within the filament, and the filament ends have distinct biochemical properties. Because the asymmetric ends of the actin filament have a distinct affinity for polymerization. Actin subunits preferentially assemble at one end, known as the barbed end and dissociate from the non-preferred end, known as the pointed end, following ATP hydrolysis and phosphate release. The subunits flow across the filament network. Only when the rates of assembly and disassembly are equal can filaments remain constant in length. Thus, the rates of monomer addition and dissociation regulate the filament length as well as the pace at which actin subunits traverse along the filament. In eukaryotes, many accessory proteins have been found to have an impact on the construction, disassembly, and reorganization of actin filaments in vivo.

2.7 MRE PROTEINS

It was known that a group of genes called mre (murein cluster e) played a significant role in determining the cell shape of many more complex-shaped bacteria, such as *C. crescentus* and the rod-shaped *Bacillus subtilis* and *E. coli*. These genes are mostly absent from coccoid bacteria and some rod-shaped bacteria, but they are also present in certain mollicutes and archaea. The mreB, mreC, and mreD genes are found in the mre cluster. Certain organisms, like *B. subtilis*, also possess mrebl (mreB-like) and mrebh (mreB homolog), in addition to many mreB-related genes. Since mutations or depletions of MreB and MreB-like proteins produce severe abnormalities in normal cell morphology, which typically lead to cell death, mreB appears to be essential. Cell size anomalies, especially in cell width, are a typical characteristic of strains depleted of MreB and mreB mutants. The morphologies of *B. subtilis* mbl mutant strains are curved, and deformed. In vivo, MreB and the other MreB-like proteins form a helical filament near the cytoplasmic face of the membrane, supporting cytoskeletal involvement. Many experiments have led to the theory that the MreB family of proteins is crucial for localizing the machinery involved in cell wall formation throughout the length of the lateral cell wall. Early research in *B. subtilis* employing a fluorescent vancomycin derivative that binds to newly synthesized peptidoglycan showed that new peptidoglycan is inserted along the cell length in a helical pattern. The vancomycin localization pattern was reliant on the Mbl protein, a MreB-like protein(Heichlinger et al., 2011). Penicillin-binding proteins (PBP-2), a component of the peptidoglycan synthesis machinery, was tested for its dependence on MreB in localization in *C. crescentus*. Similarly, the cell wall hydrolase enzyme LytE in *B. subtilis* has a MreBH-dependent pattern of localization. X-ray crystallography was used to solve the three-dimensional molecular structure of MreB, and the results showed that MreB and actin have the same fold. One obvious distinction between MreB and F-actin is the degree of axial rotation, or twist, that occurs inside these double filaments; in F-actin, the filaments twist around one another. MreB and MreB-like filaments are quite dynamic in vivo. Fluorescently labeled MreB in *E. coli*, *C. crescentus*, and *B. subtilis*, as well as MreB, Mbl, and ReBH in *B. subtilis*, have all been shown to exhibit a range of helical localization patterns with distinct filaments within a population of cells by time-lapse imaging. This is probably connected to the finding that these protein localizations vary dynamically during the cell cycle. In *C. crescentus* cells that are synchronized, for instance, it has been seen that MreB localizes as a helical structure that runs the length of the cell and collapses into a single band at mid-cell at a time that coincides with Z-ring formation. The MreB then spreads as cell division proceeds, eventually extending the length of both daughter cells. Mbl filaments in *B. subtilis* undergo continual remodeling, according to FRAP tests, and their half-lives are roughly eight minutes. This method also showed that Mbl turnover occurs along the length of the filaments and that, in contrast to F-actin, Mbl filaments do not appear to be polar. There are roughly ten times as many Mbl molecules in a cell as would be needed to assemble a single helical protofilament. It has been proposed that Mbl could assemble into short protofilaments that can bundle together into a multistranded cable without having uniform polarity. The dynamic nature of this type of cable could then result from the constant replacement of Mbl subunits inside the cytoplasm by those that dissociate from the ends of each short protofilament in the bundle.

Studies on MreB filaments conducted *in vitro* have revealed the presence of bundles of filaments, tiny ring-like structures that are too narrow. Similar to the filament morphologies seen in FtsZ, the biological significance of these structures is unknown, and some of them are likely artifacts. It is interesting to note that MreB bundled filaments exhibit improvements in stiffness, which could be crucial if MreB cables perform a mechanical function in vivo.

Additionally, it is feasible that auxiliary proteins control MreB assembly in vivo through mediating MreB–membrane or MreB–MreB interactions. As with F-actin, MreB filament development is nucleotide-dependent, with ATP and GTP stimulating filament formation. Compared to actin, MreB has a 100-fold lower critical concentration for assembly and lacks a distinct nucleation process(Carballido-López, 2006). Actin filaments function as tracks that motor proteins can use in eukaryotes. Furthermore, a wide range of proteins are known to interact with actin. No potential prokaryotic motor proteins that interact with MreB have been found. Since ParM, also known as StbA, is not necessary for regular cell activity, it is a specialized cytoskeletal element that is only needed for R1 low-copy number plasmids in *E. coli*. The par system, which consists of three components that are sufficient to transfer plasmids to opposing ends of the cell quickly after replication, actively partitions R1 plasmids. ParM, ParR, and parC make up the par system. ParM forms dynamic filaments that align along the length of the cell when it is in vivo. The R1 plasmids localize to opposing ends of the ParM filament structures, as shown by Dual-labeling immunofluorescence microscopy investigations. ParR is necessary for the ParM filament to associate with R1 plasmids. The cis-acting centromere-like parC DNA sequence expressed on the R1 plasmid is bound by the ParR protein in a specific manner.

Early sequence studies revealed that ParM shared similarities with the actin family. The ATPase domains of actin were conserved when the structure was discovered in 2002, however, the domains IA, IB, and IIB differ in that they have expanded loops, missing helices, and strand insertions. Despite sharing structural similarities with actin, ParM filaments are two-stranded and wind helically around one another. However, the filaments differ in the frequency of their helical repeats; an actin repeat occurs every 360 Å, while a ParM filament full turn occurs every 300 Å(Orlova et al., 2007). When ParM and actin are compared further, it was observed that the large regions of variation correspond to the actin interaction faces for protofilament interactions, indicating that ParM may have different assembly characteristics. ParM differs greatly from actin due to some biochemical characteristics. Unlike actin, which requires specific nucleation components for in vivo assembly, ParM exhibits rapid self-assembly, with a nucleation rate 300 times quicker than actin. PARM filaments assemble bidirectionally in a symmetrical manner but deconstruct unidirectionally, in contrast to actin unidirectional assembly. Although the ADP form is exceedingly unstable, the filaments themselves can self-assemble with a dependence on Mg^{2+} and either ATP, ADP, or one of the non-hydrolyzable ATP analogs, ATP-γ-S or AMPPNP. Actin displays a steady-state thread-milling behavior, while ParM filaments display dynamic instability. ParM is highly expressed (15000–18000 molecules/cell), although the exact molecular arrangement of the ParM filament in vivo is unknown. This suggests that either the functional in vivo ParM filament is likely to consist of multiple subunits or that filament nucleation and polymerization are dependent on the high protein concentration. The three parts (Par system) can self-assemble in vitro to form a device that can exert mechanical force appropriate for plasmid partition. It was shown that ParM filaments self-assembled into filaments, starting at ParR/parC complexes. Dynamic instability was seen in the ParM filaments, which grow and shrink quickly but only when free ends are unbound from ParR/parC complexes. This suggests that the ParR/parC complex stabilizes ParM filaments. A stable spindle that was shielded from disintegration was created when ParM filaments were bivalently coupled to ParR/parC complexes at each end.

The mechanism by which ParM extends the filaments while the ParR/parC complex is connected remains unclear because the addition of ParM monomers unexpectedly occurred at the extremities/ends of the filaments rather than in the middle. Intermediate filaments are abundant, relatively stable filaments that play a function in providing mechanical support in a variety of eukaryotic cell types. They get their name from the fact that they form filaments with a diameter between that of F-actin and microtubules. They cannot be treadmilled, they play no roles in cell motility, and no motor proteins that use IFs for tracking are known to exist. Rather than assuming transient structural responsibilities, IFs often mediate connections between cells in a matrix, however, they do display certain dynamic characteristics. The proteins that make up IFs are α-helical. The proteins in this family share a coiled motif with different N and C termini, giving them structural similarities.

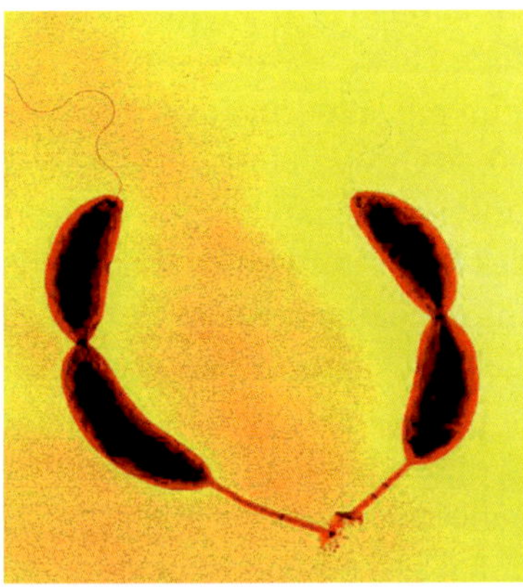

Figure 2.3 *C. crescentus*

Source: https://upload.wikimedia.org/wikipedia/commons/c/c8/Caulobacter_crescentus.jpg,*Creative Commons License.*

IF proteins exhibit significant sequence divergence and molecular weight variation. Furthermore, IF proteins have not been shown to contain any distinctive conserved sequence motifs that have made it easier to identify tubulin and actin homologs. IF proteins, on the other hand, are primarily made up of coiled sections, a feature shared by many other proteins. It is unclear exactly how they accomplish this, although filaments made of IF proteins may give the cell structural constraints appropriate for determining cell shape, as is the case with eukaryotic cells. Crescentin is the sole bacterial protein that has been assigned to the IF family thus far. Studies in *C. crescentus* provided an indication regarding the regulation of cell shape during the search for insertion mutants that alter cell morphology. When the creS gene is disrupted, the vibrioid-shaped bacteria *C. crescentus* (Figure 2.3) has been shown to generate rod-shaped cells. Crescentin, the gene product of creS, is at least believed to be a functional homolog of IFs in eukaryotes(Kelemen, 2017).

2.8 CRESCENTIN

The amino acid sequence of Crescentin is made up of heptad repeats, which are structural repeats made up of seven amino acids with four amino acids at one position, hydrophobic residues at the other, and polar residues at all other positions. This pattern is consistent with coiled structures and the IF family of proteins. The coiled-coil repeat is a weak criterion for assessing homology, and the IF proteins lack any known enzymatic or nucleotide-binding activity that would indicate their function. IFs are believed to be associated with crescentin for several reasons. Crescentin can form filaments in vitro in environments that are extremely similar to those needed for interferon assembly. These filaments do not need cofactors or nucleotides to be stable. Moreover, researchers report that the crescentin physical dimensions are very similar to those of IF filaments. Additional evidence is provided by in vivo data, which show that crescentin forms a filament that localizes to the concave side of the cell and has a lengthy helical turn. C. *crescentus* cells that have been cultured in the same medium for an extended period display a helical twist shape that is comparable in pitch to the twist seen in the crescentin filaments that are generated in vitro. It has been extremely difficult to solve the molecular three-dimensional structures of IF proteins. Moreover, neither crescentin nor IF proteins have any known structural information (molecular level), making it challenging to determine whether or not these proteins have similar protein folds. The presence of the peptidoglycan cell wall has long been associated with the integrity of bacterial cells. Researchers found that the turgor pressure of bacteria is higher than 4 atmospheres (atm). Additional research suggested that hypoosmotic shock could quickly liberate/release the amino acids that bacteria hold.

2.9 PEPTIDOGLYCANS AND OTHER DETAILS

The main structural element of the bacterial cell wall, peptidoglycan (PTG), gives the cells their shape(Cabeen et al., 2011). This has been known for more than 50 years, and pictures of isolated cell walls taken with an electron microscope provide ample evidence for this. Even after the cytoplasm and membranes have been broken down by enzymes and detergents, the bacillary organism profile is still present. PTG is a three-dimensional structure, a hollow body that surrounds the bacterial cell. More than ten times the force required to break a lipid bilayer, like the cytoplasmic membrane, is maintained by bacterial cells in the form of turgor pressure from the cytoplasm. Escherichia coli has a turgor pressure of about 4 atm, but the lipid bilayer can rupture at just about 0.2 atm. However, the cytoplasmic membrane is the target of this turgor pressure force, and the PTG is ultimately responsible for the integrity and survival of the cell. The fact that the most effective class of antibiotics, penicillins and cephalosporins, target bacterial PTG cross-linking is one of the best examples of this phenomenon. There is still debate regarding the molecular structure of PTG, uncertainty surrounds the structural components that lead to cell integrity, and more needs to be understood about the actual "in-cell" dynamics that enable the cell to withstand significant environmental changes. Although a significant amount of the enzymology about bacterial PTG production has been clarified, many issues remain. Specifically, little is known about the mechanisms governing the distribution of enzymes to certain sites and the systems governing their activity. The identification of mechanosensitive (MS) channels and their function in cell physiology provided a new opportunity for starting to comprehend cell wall formation and dynamics. Multiple kinds of MS channels were found in *E. coli* as a result of the application of patch clamping (electrophysiological technique) designed to comprehend the characteristics of ion channels in mammalian cells, to bacterial protoplasts.

2.10 MS CHANNELS

Similar channels were found in Gram-positive bacteria and the archaea. The characteristics and importance of MS channels have been more apparent in recent years. The crucial finding was that when subjected to a sharp drop in external osmolarity (hypoosmotic shock), mutant cells devoid of the channels frequently lyse. In these tests, which are similar to what happens to a bacterium in a dry area suddenly coming into contact with rainwater, the turgor pressure rises three to four times faster than it would normally in a matter of seconds. Accordingly, the pressure is currently around thirty times higher than what is required to cause the cytoplasmic membrane to lyse. Although it can offer some resistance, the cell wall is insufficient to endure these forces on its own. MS channels gate in response to an increase in membrane tension brought on by an increase in cell turgor(Cosgrove & Hedrich, 1991). This momentarily creates holes in the membrane that allow hydrated solutes to escape from the cytoplasm. This lessens the rise in cell turgor and decreases the osmotic gradient. In hypoosmotic shock, mutants lacking MS channels lyse in response to significant turgor upsurges. Thus, some bacterial cells need to have their MS channels activated in order to maintain their structural integrity. MscL and MscS from *E. coli* are the best examples of MS channels(Balleza & Gómez-Lagunas, 2009). Models for the structural rearrangements that take place during MS channel gating have been developed due to the availability of crystal structures, electrophysiological techniques, and biochemical and genetic processes. A consistent mechanism for coping with changes in external osmolarity has evolved in almost all bacterial and archaeal cells. Each organism has a different setup, especially when it comes to the kind of solutes that build up in reaction to excessive osmolarity. Due to water entering the cell through the osmotic gradient, all bacterial cells have a turgor pressure (outwards) that they must sustain. Certain solutes build up in the cell to extremely high concentrations in comparison to the surroundings, which causes the osmotic gradient. This is partially explained by the requirement for metabolic intermediates to build up in the cytoplasm to levels appropriate for rapid enzyme activity. Even when growing at a low osmolarity, studies have revealed that *E. coli* accumulates osmotically active anions to approximately 200 mM. Anions cannot build up in the cytoplasm until the proton they are accompanied by is replaced by another cation, typically K^+. Because of the extreme acidity of the cytoplasm brought on by the accumulation of 200 mM protons, DNA damage and protein denaturation would result. The elevated cytoplasmic ion concentrations, however, cannot be explained only by metabolic anions.

2.11 CELL VOLUME

Cell volume must increase for cells to grow. To facilitate the biosynthesis of new proteins, the cell must expand somewhat, which is accomplished by coordinating the synthesis of phospholipids to increase the membrane surface area. Water entering through the osmotic gradient is what is causing this expansion. To enable the controlled buildup of cations, bacteria have highly regulated K^+ transporters. K^+ ions are mostly impermeable to *E. coli* cells. This is supported by the finding that 40 mM concentrations of K^+ in the environment are necessary for the survival of the growth of mutants deficient in the three main K^+ transport systems (Kdp, Trk, and Kup). On the other hand, with even 10μM K^+, strains with all three transport systems can develop quickly. The remarkable ability of cells to hold onto K^+ is further demonstrated by their ability to maintain the cytoplasmic pool at 300 mM for extended periods of time in K^+-free solutions. This suggests that the cell membrane is generally impermeable to this cation. The cell membrane can actually withstand K^+ gradients that are almost 106 times stronger. Compared to ions, water can pass through the membrane readily. An initial loss of water from the cytoplasm brought on by the shift in the osmotic gradient is the reaction to an increase in external osmolarity. In response, *E. coli* cells take up K^+ by either triggering their highly controlled K^+-scavenging K^+-specific ATPase system, Kdp, or activating their constitutive Trk K^+ transporter. Only in low external K^+ environments is the Kdp system, a high-affinity transport system, induced or derepressed. K^+ accumulation is tightly regulated, causing the cytoplasmic K+ concentration to rise in tandem with changes in the external osmolarity. Although some net proton extrusion occurs along with the buildup of K+, increasing the cytoplasmic pH, glutamate synthesis or transport is mostly responsible for the cation balancing(Boron, 2004). Enzyme action is hampered at extremely high external osmolarities due to the buildup of K^+ and glutamate in the cytoplasm. As a result, cells have developed an additional tactic that permits quick growth at high osmolarity without compromising the turgor necessary for cell expansion.

Compatible solutes were described as substances that, when given to enzymes cultured in high salt, restored the activity of the enzymes. These solutes can alter the degree of protein hydration. A two-stage adaptation to high external osmolarity is seen in *E. coli*. Cells first take in K+ glutamate, but gradually they swap these ions out for other suitable solutes. For *E. coli* and many other bacteria, betaine, proline, and ectoine are effective solutes. Furthermore, trehalose, a disaccharide, can be synthesized by *E. coli* and some other bacteria, allowing them to utilize it as a compatible solute. Both at the level of the transport systems and through control over how the transport systems express themselves, compatible solute accumulation is firmly managed. When a cell experiences osmotic stress, it produces an accumulation of betaine, proline, and ectoine. The cytoplasmic concentrations of these substances show the external osmolarity and rise in proportion to the increase in external osmolarity. Compatible solutes displace K+ glutamate from the cell, allowing for the buildup of these solutes at a relatively constant cell turgor. Thus, the favorable effects of these solutes on protein folding along with the decrease of the inhibitory ion pools combine to produce growth stimulation by compatible solutes. The observed rapid depletion of amino acid pools during hypoosmotic shock provided the first indication that cells might have mechanisms to release solutes non-discriminately. Bacteria that have grown, or been cultured, at high osmolarity build up large amounts of solutes in their cytoplasm as a means of producing turgor. The permeability of the cell membrane is selective; it has a high permeability to water and a low passive permeability to solutes. Because of the relative permeabilities, water enters cells that are acclimated to high osmolarity conditions more quickly than any solute does during the early phase of the reaction. The membrane-restricted expansive capacity prevents the cell from expanding immediately after the inflow of water.

2.12 SELECTIVITY

Selectivity between distinct solutes is almost impossible due to the huge diameter and often hydrophilic nature of the lining of the open channels, except size. Therefore, most of the MS channels that have been investigated can easily handle any cytoplasmic molecule that is smaller than 300 to 400 Da, while the smaller channels may restrict solutes that can be discharged(Martinac et al., 2008). In a hydrated condition, the cell can lose amino acids, and organic acids that are metabolic intermediates, cofactors, disaccharides, and nucleotides very quickly. Hydrated solutes must flow through the channel pore to accomplish quick solute release because the removal of water that is hydrogen-bound to organic molecules and ions is a slow process.

On the other hand, many higher organisms also have ion-specific MS channels. These channels are thought to be more closely related to cell signaling than to the alleviation of osmotic stress because molecular groups inside the pore replace the water around the ions.

Rapid solute release removes the osmotic gradient that is causing water ingress and reduces the strain on the cell wall. Opening these channels instantly results in a variety of stressors, including loss of cytoplasmic homeostasis (pH, cation and anion balance), loss of nutrients, cofactors, and intermediates, and modification of membrane potential. The cell can withstand brief fluctuations in pH, but it will become unviable in an excessively acidic environment. Opening MS channels is therefore a very dangerous approach used solely to handle extremely high levels of stress that jeopardize the cell's structural integrity. Different bacterial MS channels have been identified using electrophysiological techniques. These channels frequently show large conductance values, more than 0.1 nanosiemens.

We may calculate the potential pores formed by open MS channels to have a diameter of several angstroms based on these conductance data. Because porin pores frequently are very asymmetric, they can function both selectively and with lower ion conductance than would be predicted given the diameters seen in crystal structures. However, during transmission across the membrane, minute compounds like carbohydrates and antibiotics can momentarily block the pore. For the solutes to pass through the pore without being restricted, the MS channels need to be much bigger. MscL and MscS from *E. coli* were well-researched. Comprehensive electrophysiological analysis, in conjunction with molecular genetics, biochemistry, and biophysics, has led to an understanding of their properties.

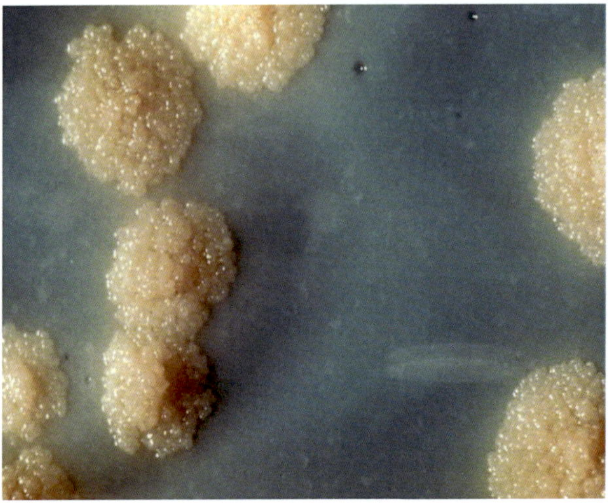

Figure 2.4 Mycobacterium TB

Source: https://upload.wikimedia.org/wikipedia/commons/0/0a/TB_Culture.jpg,*Creative Commons License.*

Mycobacterium TB (Figure 2.4) and *E. coli* provided the crystal structures for MscL and MscS, respectively(Steinbacher et al., 2007). Although both proteins perform the same job of forming large holes, they differ in terms of size, complexity, and mechanism. The exact location of the MS channel proteins was formerly unknown, but it is now known that they are parts of the cytoplasmic membrane.

But through its long amino peripheral domain, which has a periplasmic position, one class of MS channel protein, the MscK family, can interact with both periplasmic proteins and those in the outer membrane. The transmission of the stress signal that gates these channels is entirely communicated through the lipid bilayer; no accessory proteins are necessary for MS channel activation. Bacteria have MscL channels.

MscL is a homopentamer made up of subunits that are roughly 17–20 kDa and typically have 120–156 amino acid residues. Several higher organisms have proteins with sequences that are very similar to the transmembrane (TM) helices of MscL. There is currently no information indicating that archaea have MscL homologs in their genomes. At almost three nanosiemens, the open MscL channel has the greatest conductance of any channel that is currently known. The channel in the *E. coli* membrane opens at pressures higher than 150 mmHg (1 atm = 760 mmHg), and this pressure range is in close proximity to the membrane lytic pressure. Each channel opening lasts only a few milliseconds and occurs within three microseconds of the pressure shift. If the pressure stays over the gating threshold, the channel alternates between closed and open states. Few of these channels are known to exist in the cytoplasmic membrane of bacterial cells. Every MscL subunit is made up of two TM domains, called TM1 and TM2, that are bordered by carboxy- and amino-terminal α-helices and divided by a periplasmic loop of varying length. About 80% of each TM α-helix length, or thirty residues, is incorporated in the membrane bilayer.

Chapter **3**

LACTIC ACID BACTERIA

Contents

3.1 INTRODUCTION

A diverse class of Gram-positive bacteria known as lactic acid bacteria generate lactic acid as a primary byproduct of their fermentative metabolism. The primary genera of lactic acid bacteria are *Lactobacillus*, *Lactococcus*, *Streptococcus*, *Pediococcus*, *Enterococcus*, *Leuconostoc,* and *Carnobacterium*. One of the most basic methods of food preservation is lactic acid fermentation, although lactic acid bacteria (LAB) only start to damage our food when the shelf life is extended. Lactic acid fermentation, for instance, may contribute to the development of some textures, enhance certain distinctive sensory qualities, and even provide advantages that support and enhance human health. It has long been thought that LAB carries out those tasks through a variety of metabolites. For this reason, the food and pharmaceutical industries are paying close attention to their potential application as cell factories for desirable microbial metabolites. The use of LAB metabolic and genetic engineering encourages the development of both primary and secondary metabolisms, which are more complicated. A food flavor is a crucial component that greatly influences consumer acceptance. Microorganisms in the starter cultures perform several biochemical transformations of milk components to create the tastes of dairy fermentations, such as cheeses and yogurts. The metabolism of fat (lipolysis), protein (proteolysis), and carbohydrates (glycolysis) are the three primary metabolic processes.

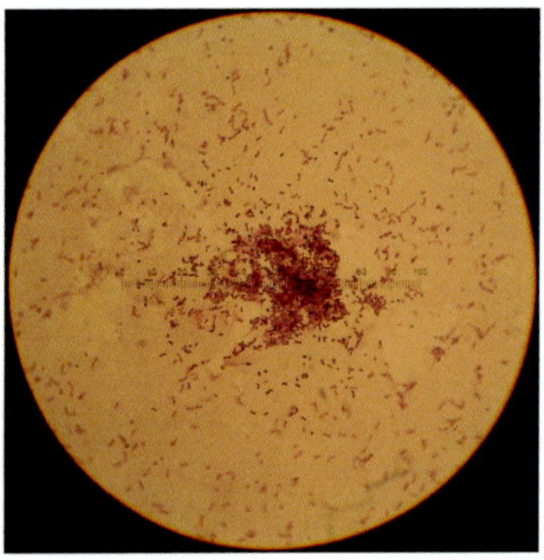

Figure 3.1 *Lactococcus lactis*

Source: https://upload.wikimedia.org/wikipedia/commons/4/4f/Lactococcus_lactis.jpg,*Creative Commons License.*

Lactic acid bacteria (LAB), such as *Lactococcus lactis*, (Figure 3.1) Lactobacillus species, *Streptococcus thermophilus*, Bifidobacterium species, and Leuconostoc species, are the main microorganisms in these starter cultures of dairy fermentations(Liu et al., 2014). The principal source of carbon for LAB in fermented dairy products is lactose. For LAB, there are two fermentation types: heterofermentation and homofermentation. They produce distinct lactose fermentation byproducts. Homofermentation mostly produces lactic acid, while heterofermentation produces ethanol, carbon dioxide, and acetic acid, among other byproducts. It is interesting to note that a homofermentation can transition into a mixed-acid metabolism under specific conditions, such as carbon limitation or aerobic culture, which yields a variety of metabolites, including many aroma molecules like acetaldehyde and diacetyl. Casein proteolysis, which consists of two stages: amino acid metabolism and proteolysis, is crucial to the development of flavor in dairy fermentation. Casein is broken down into oligopeptides by cell-envelope proteinases (CEPs), and then peptidases break it down into amino acids. Subsequently, free amino acids can be converted into a variety of flavoring substances, including alcohols, amines, ammonia, phenols, aldehydes, and indole. The main building blocks for flavor generation are branched-chain amino acids (valine, leucine, isoleucine), aromatic amino acids (phenylalanine, tyrosine, tryptophan), and sulfuric amino acids (cysteine and methionine). Some of these sulfur compounds have off-flavors for yogurt, even if they are crucial for cheese flavors.

3.2 AROMATIC COMPOUNDS

In dairy fermentation, the formation of aroma compounds also depends on the lipolysis of fat. It has been shown that lipolysis has a minor effect on yogurt flavor development but a major impact on the flavor development of long-ripened cheeses. Lipase breaks down lipids into fatty acids (FAs). The unsaturated FAs undergo oxidation to produce hydroperoxides, which quickly change into hexanal or unsaturated aldehydes. To create cyclic molecules, unsaturated FAs can also transform into four or five hydroxy acids. Dairy products mostly contain γ- and δ-lactones, which are cyclic chemicals that provide fruity aromas. Certain lipolysis and lipid oxidation can also provide an off flavor. For instance, the oxidized flavors in dairy products are caused by aldehydes and ketones that are formed during lipid oxidation. LAB is also rich in esterases that can directly produce taste esters from glycerides and alcohols through an alcoholysis reaction. To create ethyl butanoate and ethyl hexanoate, for instance, LAB can esterify ethanol with butyric acid and hexanoic acid(Abeijón Mukdsi et al., 2009). It is interesting to note that flavor esters are formed more often from diglycerides and monoglycerides than from triglycerides. Another crucial flavor component in dairy products is acetaldehyde, a colorless liquid with the molecular formula CH_3CHO that has a green apple flavor. In yogurts, *Lactobacillus bulgaricus* and *Streptococcus thermophilus* both produce acetaldehyde. A crucial aroma component for yogurt flavor is acetaldehyde. Acetaldehyde levels in the range of

23–40 mg/kg are necessary for flavored yogurt. Pyruvate is the main metabolic substrate for acetaldehyde biosynthesis, and it is converted to acetaldehyde by aldehyde dehydrogenase or α-carboxylase.

3.3 DIACETYL

Butanedione, sometimes known as diacetyl, is a yellow or green liquid having the chemical formula $(CH_3CO)_2$. It can provide a variety of fermented milk products, like cheese, cream, yogurt, etc. Furthermore, diacetyl is employed as a flavoring additive in non-milk products such as baked goods, candies, soft drinks, and other items in the food business. Nonetheless, diacetyl is a significant cause of off-flavor in wine and beer.

Diacetyl is a byproduct of the fermentation of lactic acid, particularly by *Lactococcus lactis* and/or Leuconostoc species from the metabolism of citrate. Pyruvate produced by citrate metabolism may accumulate as α-acetolactate, the precursor to diacetyl via an oxidative decarboxylation process. Citric acid in fresh milk can be readily converted to pyruvic acid without the need for nicotinamide adenine dinucleotide (NAD). Diacetyl synthesis in *Lactococcus lactis* fermentation is dependent on several external variables/factors, including temperature, pH, oxygen content, culture medium, and the kind and proportion of carbon and nitrogen sources. Research has shown that *Lactococcus lactis* grown in aerobic environments produced more diacetyl than it did in anaerobic environments. Citric acid added to the medium can increase the production of diacetyl in anaerobic conditions. Diacetyl reductase activity could be effectively inhibited to preserve diacetyl by cooling the products as soon as they reached the desired acidity and flavor. Many important enzymes involved in fermentation, such as citrate lyase, oxaloacetic acid decarboxylase, alpha acetyl synthetase, NADH oxidase, alpha acetyl lactic acid decarboxylase, diacetyl reductase, and others, are strongly linked to the formation of diacetyl. These enzymes are made using molecular biology techniques; inactivation or overexpression can lead to an increase in intermediate accumulation, such as pyruvic acid and alpha acetyl lactic acid, which can alter the metabolic pathways that lead to diacetyl and enhance the production of diacetyl. Esters comprise a vast array of volatile chemicals that are typically found in fermented foods at levels that surpass the threshold of sensory perception. They are frequently ethyl esters of the C_2–C_{10} straight-chain fatty acids and help cheeses acquire their distinctive fruity aromas. Different free fatty acids (FFA) may be produced as a result of lipolysis during the preparation of cheese, and these products have an impact on its flavor. Esters can mitigate the effects of off-flavors brought on by excessive concentrations of short-chain FFA(Bicas & Rodriguez-Amaya, 2021). Although the role of lactic acid bacteria, including lactobacilli, lactococci, enterococci, and streptococci, in the esterase/lipase system in the synthesis of taste ester compounds has been well-documented, the underlying process is still mostly unclear.

3.4 GLYCOLYSIS

Glycolysis, also known as the Embden-Meyerhof-Parnas (EMP) pathway, (Figure 3.2) is the first step in the conversion of glucose to pyruvic acid(Sánchez-Pascuala et al., 2017). Pyruvic acid is later converted to lactic acid. Hence, the only byproduct of the homofermentative pathway is lactic acid, with a maximum potential production of 2 mol lactate for every 1 mol glucose.

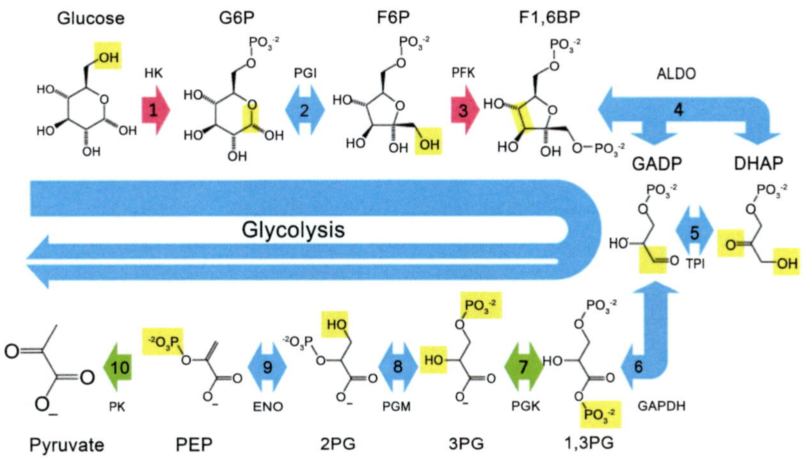

Figure 3.2 Glycolysis

Source: https://upload.wikimedia.org/wikipedia/commons/c/c0/Glycolysis_metabolic_pathway_3_annotated.svg,*Creative Commons License.*

The microorganisms *Lactobacillusbulgaricus*, *Lactobacillus helveticus*, *Lactobacillus acidophilus*, and *Lactobacillus amylophilus* use this pathway for glucose metabolism. On the other hand, glucose is metabolized via the phosphoketolase (PK) pathway in a heterofermentative method.

In addition to lactic acid, this process also produces CO_2, ethanol, and/or acetic acid. One of the important byproducts of LAB is lactic acid (LA), which is needed due to its many uses in the food, pharmaceutical, cosmetic, and chemical sectors. An important application of lactic acid is in the food industry, where it is used to make cheese and yogurt. *Streptococcus thermophilus* and *Lactobacillus bulgaricus* co-ferment together to produce the majority of the lactic acid used in yogurt production. Lactic acid lowers the pH of cheese during the production process, which encourages the aggregation of casein micelles. Furthermore, lactic acid and its salts are frequently used as food additives and are regarded by the Food and Drug Administration or FDA as generally recognized as safe (GRAS)(Plavec & Berlec, 2020). Hence, lactic acid can be added directly to various food products, which has the dual benefits of preventing the growth of unwanted microbes and adjusting flavor to form the desired sensory characteristics. Furthermore, during the grain-producing process, lactic acid is also produced spontaneously.

3.5 POLYMERS

Propylene glycol and acrylic polymers, among other small and large molecules, can be produced with lactic acid as a raw material. Their polymers can be produced into industrial products for labeling, packaging, and prosthetic devices because of their excellent biodegradable and biocompatible qualities. For instance, the textile and pharmaceutical sectors make extensive use of polylactic acid. One of the important byproducts of LAB is acetic acid, which is produced by the heterofermentative pathway of glucose in LAB by bacteria such *Lactobacillus brevis*, Lactoferrones fermentum, and Bifidobacterium. Strong bacteriostatic activity and bacteriostasis are characteristics of acetic acid. Bacteria, fungi, yeast, and other potentially dangerous microbes can all be prevented from growing and developing by acetic acid.

The bacteriostatic effect is caused by acetic acid and other organic acids generated by LAB, which can lower pH, damage Gram-negative bacterial membranes, and obstruct the development of bacterial enzymes. As a result, food preservation uses acetic acid and other organic acids generated by LAB extensively. Acetic acid is good for human health. Acetate, via G protein-coupled receptors (GPCRs), particularly GPR43/FFAR2, could be crucial in preserving host homeostasis(Tan et al., 2017). Acetate has been linked to enhanced glycemic management and weight loss.

Acetate supplementation has been shown in rodent and human studies to enhance glucose tolerance. The fundamental process involves the activation of GPR43 by acetate, which leads to the production of enteroendocrine hormones that influence intestinal transit. These hormones include peptide YY (PYY) and glucagon-like peptide-1 (GLP-1). Additionally, the increase of the intestinal gluconeogenesis (IGN) effect is a result of acetate activating GPR43. By activating GPR43 in white adipose tissue (WAT), which is advantageous to adipose metabolism, acetate supplementation may also help lower obesity. Acetic acid helps relieve constipation by promoting the large intestine peristalsis. Acetic acid may prevent influenza virus infection and the growth of Ehec O157 in the gastrointestinal tract.

3.6 ACETIC ACID AND GABA

Acetic acid functions in multiple ways in the host brain. A minimal quantity of acetate can penetrate the blood-brain barrier (BBB) and subsequently trigger hypothalamus neurons, so inducing satiety. Acetate activates GPR43, which is essential for microglia development in the brain. Moreover, acetate may alter the BBB permeability. Being a non-protein amino acid, gamma-aminobutyric acid (GABA) is a well-known primary inhibitory neurotransmitter in the central nervous system. Although GABA is widely distributed in both plants and mammals, its natural concentration is low. Thus, the majority of GABA is made chemically or through the use of microbes for bioconversion. However, the food industry is not allowed to use chemically generated GABA. As a result, one important way to produce GABA or GABA-rich meals is through the bioconversion of GABA utilizing food-grade microbes, particularly LAB. It has been established that several LAB species and subspecies that have been isolated from diverse habitats may generate GABA. It demonstrates how different LABs have varying capacities for producing GABA. Fermented foods have high GABA content. Glutamate decarboxylase or GAD, which catalyzes the decarboxylation of L-glutamate to create GABA, is recognized to be the enzyme responsible for producing GABA. To preserve intracellular pH equilibrium in acidic environments, the LAB with high GABA contents may remove intracellular protons during the GAD catalytic process. The majority of LAB with high GABA productivity have been reported to belong to Lb. brevis and Lb. Plantarum. GABA has many physiological effects, such as those that are antidepressant, anti-anxiety, and antihypertensive. GABA or meals high in GABA generated by LAB also have anti-depressive and antihypertensive properties for the host(Rashmi et al., 2018).

3.7 LINOLEIC ACID

A class of geometric and positional isomers of linoleic acid (LA) with conjugated double bonds, known as conjugated linoleic acids (CLAs), are mostly found in milk and meat products(Yaquoob et al., 2006). The common isomers of CLA are trans-10, cis-12-CLA (10-CLA), and trans-9, trans-11-CLA (9-CLA) and they are mostly produced from linoleic acid. The microbial synthesis of CLA could be a viable technique for industrial production. Many bacteria have been shown to convert free LA into CLA, although some lactic acid bacteria (LAB) strains, such as Bifidobacterium, Lactobacillus, and Streptococcus, have garnered more attention than other LAB strains that produce CLA because of their positive health effects. Due to its health-promoting properties, such as its anti-obesity, anti-carcinogenesis, anti-inflammatory, and antidiabetic effects, CLA has drawn a lot of attention lately. Many studies have indicated that its ability to prevent obesity may be attributed to a variety of mechanisms. CLA may primarily bind to PPARγ to regulate preadipocyte development. CLA may reduce energy or food intake, increase energy expenditure, decrease preadipocyte differentiation and proliferation, decrease lipogenesis, and promote lipolysis and fat oxidation. Moreover, CLA may reduce leptin, boost adipocyte death, and raise tumor necrosis factor-α (TNFα) levels to achieve its anti-obesity action. CLA exhibited strong inhibitory effects on skin, stomach, breast, or colon cancers in mice and rats. It is interesting to note that in vivo human research also confirms that CLA has strong antiproliferative effects on breast and prostate cancers.

It has been proposed that the anti-carcinogenesis properties stem from its ability to inhibit tumor growth and metastasis. Both benign and malignant tumors can be quickly suppressed by CLA. The 10-CLA isomer appears to function through regulating apoptosis and the cell cycle, whereas the 9-CLA isomer primarily affects arachidonic acid metabolism. Complex molecular mechanisms underlie CLA antiproliferative actions. But it might also be accomplished in part by decreasing the tumor lipid oxidation, prostaglandin (PG) metabolism, vitamin A levels, and cell proliferation. Many studies have demonstrated that adding CLA to the diet can successfully prevent diabetes, particularly type II diabetes. Many studies indicate that the best CLA isomer to prevent type II diabetes is 10-CLA. CLA may mainly bind with transcription factors like PPARγ to influence glucose metabolism. CLA may control the expression of genes connected to lipid and glucose metabolism. CLA sensitizes adiponectin, which will increase insulin sensitivity. Research conducted both in vitro and in vivo has shown that CLA is an efficient modulator of immune function. All of the active CLA isomers, 9- and 10-CLAs, can trigger both innate and adaptive immunological responses to modulate immunity, including the generation of eicosanoids, cytokines, and immunoglobulin.

3.8 BACTERIOCINS

Many Gram-positive (+) and Gram-negative (−) bacteria produce proteins or polypeptides with antibacterial properties, which are collectively referred to as bacteriocins. Antibiotics and bacteriocins are not the same. The primary distinction is that although antibiotics have a wide spectrum of antimicrobial activity, bacteriocins exhibit antimicrobial activity against closely related bacteria. Because most lactic acid bacteria (LAB) have the status of qualified presumption of safety (QPS) and are generally regarded as safe (GRAS), the generation of bacteriocins from LAB has received a lot of interest recently. Bacteriocins derived from LAB have potential applications in the food industry as food preservatives in medicine as therapeutic agents. LAB be classified into four classes: Class I (lantibiotics), Class II (non-lantibiotics), Class III (large heat-labile murein hydrolases), and Class IV (the lipid or carbohydrate-containing bacteriocins) based on the diversity of their structure and physicochemical property(Wu & Li, 2018). Class III bacteriocins were classified as bacteriolysins and Class IV bacteriocins as circular bacteriocins. Lanthionine and methyl-lanthionine are the main structural components of class I bacteriocins, often known as lantibiotics. They are tiny peptides, measuring less than 5 kDa. Most of the 60 or so antibiotics that have been identified recently come from LAB. They often target pathogen cell walls, particularly those of Gram-positive bacteria, to provide antibiotic effects. Among these, nisin (*Lactococcus lactis*), a Class I antibiotic, has been the subject of extensive research. Nisin is the only Class I bacteriocin that has been commercialized and used in the food business globally up to this point. Microorganisms that cause food deterioration and are highly harmful, like S. aureus and L. monocytogenes, are actively resisted by nisin. Nisin could kill specific bacteria by blocking their enzyme function, even at lower concentrations. The precise process by which it works is primarily due to nisin binding to lipid II, which stops lipid II from moving peptidoglycan subunits from the cytoplasm to the cell wall. This blocks the formation of the cell wall and results in cell death. In higher concentrations, the nisin-lipid II molecule complex may cause the target bacterial cell membrane to become perforated, damaging the physiological processes and functions of the cell and resulting in cell death. As a result, the target bacterial mortality is triggered by the binding of nisin with lipid II, which disrupts the cell membrane and prevents the production of the cell wall.

ATP-binding cassette transport systems carry class II bacteriocins, which are heat-stable, small (<10 kDa), non-lantibiotics, or antibiotics that resemble pediocins(Simons et al., 2020). Pediocin PA-1/AcH is a Class II bacteriocin that has been marketed and used in the food sector globally. Class II bacteriocins often exert their antimicrobial actions by making the target bacterial membrane more permeable. This gradually causes cytoplasmic chemicals to seep out and ultimately results in cell death. Large (>10 kDa) peptides, heat-labile lytic bacteriocins, and non-lytic bacteriocins, including lysostaphin, helveticin J, helveticin V, and zoocin A, are examples of class III bacteriocins. Lysogenic bacteriocins typically cause bacterial cell walls to rupture to exert their antibacterial properties. Due to the presence of lipid or carbohydrate moieties, several bacteriocins including plantaricin S, leuconocin S, and uberolysin are susceptible to lipolytic or glycolytic enzymes. Because LABs have a positive impact on food nutrition, organoleptic properties, and shelf life, they are widely employed in fermented foods. The food and pharmaceutical industries have recently shown interest in the bacteriocins produced by LAB. Bacteriocins can function on their own or in conjunction with other preservative techniques. Nisin and pediocin PA-1 are the only two bacteriocins that have been made available for use in the food sector. Nonetheless, LAB bacteriocins have potential applications as food preservatives. Bacteriocins can be used in food systems in a variety of ways, including direct inoculation of LAB that produces them, addition of the bacteriocin, and use of LAB-fermented products that produce them. Many studies have examined the role that bacteriocins play in food preservation, particularly in the biopreservation of dairy, egg, vegetable, and meat products. Among these, it has been demonstrated that nisin A and nisin Z are good at preventing microbial contamination of food. Class IIa bacteriocins can fight foodborne and highly pathogenic *L. monocytogenes* in cold, ready-to-eat food products. It is also possible to successfully avoid *L. monocytogenes* infection when fermented pork sausage is made utilizing the pediocin PA-1/AcH generating LAB as the starter culture.

It has been demonstrated that adding bacteriocins to food packaging films or surfaces effectively preserves food. For instance, adding pediocin PA-1/AcH to packing film may considerably lower the amount of *L. monocytogenes* on the surface of meat. The extensive use of LAB bacteriocins is currently restricted by some important issues, including their high cost of production and poor effectiveness against Gram-negative foodborne pathogens. Hurdle technology, which combines bacteriocins with other preservation techniques, could be used to address these deficiencies. Certain methicillin-resistant *Staphylococcus aureus* (MRSA) strains and vancomycin-resistant *Enterococcus faecalis* (VRE) strains are examples of multidrug-resistant (MDR) pathogens that are effectively inhibited by certain bacteriocins(Li et al., 2023). Lacticin 3147 has been shown to be effective against the MRSA and VRE strains. The fact that bacteriocins are typically inert to the normal gut microbiota is another benefit of employing them for treatment rather than traditional antibiotics. Bacteriocins are easily biosynthesized using bioengineering to either boost their activity or specify target microorganisms, in contrast to conventional antibiotics. Because of all these benefits, bacteriocins have potential uses in the pharmaceutical and therapeutic sectors. Since vitamins are known to be extremely important for maintaining human health, many efforts have been made to generate them efficiently in vivo or in vitro using a variety of techniques. Vitamins can be simply divided into soluble (VB) and insoluble (VK and VE) categories based on their varying solubilities in water.

3.9 ELECTRON CARRIERS

A necessary component of two-electron carriers (flavin mononucleotide, or FMN, and flavin adenine dinucleotide, or FAD), riboflavin is found in a variety of foods, including meat, eggs, cereals, and some vegetables. Humans are not able to synthesize. Despite the abundance of dietary resources, riboflavin deficiency still occurs and causes a range of inflammations in the lips, skin, mouth, and mucosa. This further encourages research on the mechanisms by which bacteria, particularly LAB, produce riboflavin. Researchers are becoming increasingly interested in the gene operators and enzymes involved in the synthesis of riboflavin in LAB, as the production of substantial amounts of the vitamin was typically dependent on the specific LAB species or culture conditions. When all four biosynthetic genes (ribG, ribH, ribB, and ribA) were overexpressed at the same time in *Lactococcus lactis subsp. cremoris* strain NZ9000 carrying pNZGBAH, substantial riboflavin overproduction was reported(LeBlanc et al., 2005). Interestingly, nisin was also involved in riboflavin expression. In general, the biosynthesis of riboflavin is initiated by its precursors, ribulose-5′-phosphate and GTP. A more comprehensive understanding of the process or signal involved in producing riboflavin would allow us to better understand how LAB and host interact with diet. Folic acid actively participates in several metabolic processes, such as the methyl group, amino acid, and vitamin biosynthesis, as well as DNA activities related to growth development, all of which have an impact on the human body. Folic acid cannot be produced in vivo by humans, necessitating dietary consumption. It was revealed that modified bacteria, particularly LAB, might be used to manufacture/produce folic acid in fermented foods due to cost and simplicity.

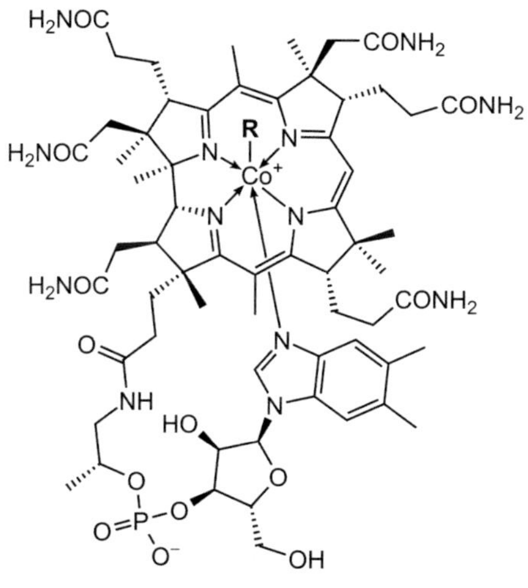

Figure 3.3 Vitamin B_{12}

Source: https://upload.wikimedia.org/wikipedia/commons/4/4c/Cobalamin_skeletal.svg,*Creative Commons License.*

This was a unique way to address the folic acid deficiency in some diets. Megalin, reduced folate carrier, and other folate-binding proteins in various organisms helped absorb folate, which was changed from polyglutamate to monoglutamate by intestine glutamyl carboxy conjugase. Specifically, the effectiveness and affinity of a carrier for folate are complex and depend on the kind, quantity, and dominance of 5-methyl-tetrahydro folate. Six-hydroxymethyl-7,8-dihydropterin pyrophosphate (DHPPP), para-aminobenzoic acid (PABA), dihydrofolate (DHF), tetrahydrofolate (THF), 7,8-dihydropteroate (DHP), and guanosine triphosphate (GTP) are all necessary for the production of folate de novo, which in turn produces folic acid. Some of these compounds function as precursors, while others supply the required energy.

3.10 VITAMINS

Four pyrrole rings and one cobalt atom make up vitamin B_{12} (cobalamin), (Figure 3.3) which is unique in that it is the only vitamin that contains a metal atom. Vitamin B_{12} has a fundamental role in human health, particularly for the treatment of severe anemia disease, was established by its involvement in two metabolic processes in vivo: the production of methionine with homocysteine assisted by 5-methyltetrahydrofolate (5-MTHF) and the formation of succinyl-CoA converted from methyl malonyl-CoA with the aid of methyl malonyl-CoA mutase(Huemer & Baumgartner, 2019). Mammals, including humans, can activate various forms of VB_{12}, but they are unable to synthesize it, which makes them dependent on food consumption. Aside from a variety of dietary sources like meat, eggs, and fish, microorganisms play a major role in the synthesis of VB_{12}; in fact, they operate as a stimulant for the revival of probiotic bacteria like LAB, which are used to produce VB_{12}. Researchers are particularly interested in the absorption and metabolism of VB_{12} since it is a valuable and finite resource in the human body. The human body is known to include two main VB_{12} receptors: the cubilin-amnionless (AMN) complex and the intrinsic factor-B12 (IF-B12) receptor. However, in serum, VB_{12} is coupled with haptocorrin (HC) and transcobalamin (TC), showing that VB_{12} is transported. Atrophic gastritis and achlorhydria were not always linked to a dysfunction in food VB_{12} absorption. However, because certain intestinal bacteria do create trace amounts of VB_{12}, some researchers are concerned about how the bacterium may affect the gut microbiome community and the way bacteria interact with their hosts. The majority of gut microbes in humans have a high probability of competing with the host for VB_{12} and pertinent cofactors, such as corrinoids. It was discovered that VB_{12} production may have an impact on aging. VB_{12} metabolism and absorption are acknowledged to be important for human health. Additional studies into the interactions between intestinal bacteria and the host on dietary VB_{12} may reveal the cause of VB_{12} malabsorption and provide a promising treatment for diseases associated with VB_{12} deficiencies.

Vitamin K is found in two forms: menaquinones (vitamin K_2), which are mostly derived from animals and gut bacteria, and phylloquinone (vitamin K_1), which is found in green vegetables(Bonaldo & Leroy, 2024). It is known to be closely associated with osteoporotic fractures and intracranial hemorrhagic disease in babies because it helps convert glutamic acid to γ-carboxyglutamic acid (Gla), which affects tissue calcification and blood coagulation through its combination with matrix Gla-protein (MGP) and osteocalcin. LAB can produce a certain quantity of menaquinone from the VK_2 resource, which has piqued the interest of the dairy sector. This is because fermented products have higher nutrition and can thus be expected to be profitable. Furthermore, menaquinone is not only produced by LAB; it also affects LAB respiration and metabolism. In particular, menaquinone and heme can stimulate respiration in some LAB species by releasing electrons, which in turn promotes growth and yield. However, considering the population of gut bacteria, there has been a new trend to use intestinal bacterial production of menaquinone as a substitution for human health status. There may be an important connection between VK2 and colitis. Specifically, the precursor of VK2, 1,4-dihydroxy-2-naphthoic acid (DHNA), can activate the aryl hydrocarbon receptor (AhR), which is known to suppress inflammatory bowel diseases (IBDs). Because VK2 is involved in the formation of microorganism membranes, researchers are currently searching for new antibiotics that specifically target bacterial membranes. One such antibiotic is lysocin E, which is effective in killing bacteria because it can affect VK2 in membranes. More research on the distribution of dietary VK2 between the host and gut flora, as well as potential applications of VK2 to biomarkers that reflect the health of the human body, will revive our understanding of how the gut absorbs nutrients.

A byproduct of many LAB O_2-dependent metabolic processes is hydrogen peroxide (H_2O_2). The H_2O_2 produced by LAB is established to have bactericidal effects on Gram-negative bacteria and bacteriostatic effects on Gram-positive bacteria(Ding et al., 2024). The environment can determine whether this feature is desired or undesirable. LAB produces H_2O_2, which has antibacterial properties in food. LAB produces H_2O_2, which is crucial for regulating vaginal health(Kang et al., 2019). Approximately 80% of vaginal strains, mostly vaginal isolates of *L. crispatus* and *L. jensenii*, have been shown to generate H_2O_2. Many studies have suggested that the presence of vaginal lactobacilli that produce H_2O_2 may protect against infection by pathogens linked to bacterial vaginosis (BV), Herpes Simplex virus type 2, and HIV-1(Kovachev, 2018). In addition, H_2O_2-producing lactobacilli may interact with superoxide anion, hydroxyl radicals, and hypochlorous acid to potentially have anti-vaginal cancer effects. Further research elucidated the mechanisms behind LAB's acidification capacity. Numerous sources focused on how resistant LAB is to physical harm in an acidic environment. There are prominent variations in the acidification capacity of certain oral bacteria. Magnesium release at pH 4.0 was shown to be similar for all streptococcal species. *Lactobacillus casei*, on the other hand, showed damage below pH 3.0 and was more resistant. The migration of protons on the *L. casei* membrane was measured. In addition, adding an ATPase inhibitor may cause *L. casei* proton permeability to increase, indicating that a proton enters the cell membrane and that proton-translocating ATPase (F_1F_0-ATPase) was engaged in the outflow process. During glycolysis, F_1F_0-ATPase can pump multiple protons out of the cytoplasm by using ATP. Many investigations have demonstrated that some F-ATPase activity in microorganisms can move protons from the cytoplasm to preserve pH homeostasis. F-ATPase in LABs may be more resistant to acid at low pH conditions. Because of its high level of acid tolerance and relative competitiveness against other species, LAB F-ATPase is important.

3.11 OTHER LAB PRODUCTS

Under some conditions, LAB can also create cyclopropane-containing fatty acids (CFAs), such as lactobacilli acid (C19cyc11) and dehydrosterculic acid (C19cyc9)(Chen & Lu, 2018). With the help of CFA synthase, methyl groups in S-adenosyl methionine can be converted to fatty acids, which in turn can create CFAs. *L. casei* ATCC334 could result in an increase in CFA and saturated fatty acid levels at the expense of UFAs. These findings suggested that the strong response in *L. casei* might involve CFA production, which is seen in *L. helveticus* and L. *sanfranciscensis*. Both *L. helveticus* and *L. sanfranciscensis* contain CFAs. relA gene is involved in *Streptococcus mutans* starvation response and its ability to produce CFAs in culture. After growing in biofilms, the relA mutant strain in this setting displayed a greater acid tolerance than the parent strain, which was one of the main impacts of the mutation. Apart from the previously established mechanisms of acid tolerance, there exist additional systems inside LAB that manage the impact of acid-mediated damage and enhance the degressive internal pH values. For instance, it is believed that the release of ammonia to raise pH is a uniformly distributed acid resistance mechanism. Ammonia sources that are commonly found include arginine and amino acid agmatine. The ammonia release mechanism involves the arginine deiminase system and is dependent on the activity of the enzymes ornithine carbamoyltransferase, arginine deiminase, and carbamate kinase. As a component of an arginine-ornithine antiporter system, arginine crosses cell membranes and aids an arginine-specific porter. When arginine is broken down by arginine deiminase, it can be transformed into citrulline and ammonia. Ornithine transcarbamylase splits citrulline to produce ornithine and carbamoyl phosphate(Aldrian et al.,

2024). When carbamoyl phosphate is converted to carbon dioxide and ammonia during substrate-level phosphorylation, carbamate kinase can break it down. These interconnected systems ultimately use a sequence of energy-formation reactions to transfer ammonia into cells and an ATP.

L-malate can be transformed into carbon dioxide through the malolactic fermentation pathway (MLF). This process involves L-malic acid releasing carbon dioxide through the action of the malolactic enzyme, forming a single molecule of lactic acid and carbon dioxide, which can then be converted into bicarbonate by carbonic anhydrase. Lactic acid was generated via an antiporter mechanism, and an L-malate molecule was employed. The MLF route was utilized in the wine sector, among other LABs, due to its widespread presence. The MLF pathway is connected to the host bacterial defensive buffering. During the fermentation of sugar, LAB may be able to thrive in the presence of alcohol and at low pH. Since different species and strains of lactic acid bacteria differ significantly even under the same growth conditions and media composition, it is difficult to determine exactly how these bacteria employ their acid tolerance. Therefore, transcriptional profiling and proteomics were used. Starvation had a specific effect on the guanidine nucleotide pools of LAB, and the Embden–Meyerhof–Parnas (EMP) system interfered with the passage of carbon. The ideal location to control cell variation, the port of conversion, carbon flow, and phosphate pools is (p)ppGpp. *Lactococcus lactis* reacts quickly to acidification involving pppGpp. More *L. lactis* mutant strains than the parent strain (MG163) had a protein deficit in relA, guaA (GMP synthase), and pstS (phosphate transporter). Examples of proteins that may overlap with the mutant strains and be relevant under acidic conditions include RecA (DNA-repair-recombination protein), pyruvate carboxylase, CTP synthase, glutamyl-tRNA synthetase, R30S ribosomal protein S1, and the DNA polymerase subunit.

3.12 PROBIOTICS

The popular probiotics, LABs, are utilized to produce wine, fermented dairy products, and foods based on vegetables(Swain et al., 2014). The most important characteristic is bile tolerance, which is generally understood to be the ability of the bacteria to proliferate and operate as probiotics. Bile typically consists of water, phospholipids, cholesterol, bile acids, and the pigment biliverdin. It results in the production of a yellow or green solution in mammalian liver pericentral hepatocytes. Certain biomolecules made from cholesterol may be eliminated by the bile. The physiological purpose of bile is to aid in the absorption of lipophilic substances from diet. Additionally, bile is essential for the development of the human gut flora. Bile generally has the potential to impair the function as a cell membrane (LAB). Research on the bile stress response in LAB and bifidobacteria has demonstrated that the integration of multilateral responses resulting from bile resistance can shield cell membrane resistance from bile acids. This process involves the destruction or restoration of proteins, the removal of oxidative stress, the facilitation of DNA repair, and the augmentation of energy generation by upregulating sugar metabolism. LAB can improve the intestinal microbial balance of the host, which has a positive effect. Consequently, LABs can be resistant to the digestive process and the enzymes. Bile travels to the duodenal region of the small intestine when bacteria enter the digestive tract, which lowers the bacterial survival rate. One element that lowers strain survival in culture conditions is bile acid. The lipid bilayer's ability to flip-flop bile acids boosts the membrane's tensile strength. Therefore, if bile acid levels are excessive, a large amount of it will be applied to the cells. The proton conductance and membrane osmosis capacity will be affected by an increase in bile acid concentration.

One crucial consideration when choosing probiotic strains is their capacity to tolerate exposure to bile. Bile contains DNA lesions, acid, oxidative, and osmotic stressors, all of which have an impact on the membrane properties of bacteria and their cells. Consequently, when researching bile tolerance, various variables such as oxidative, salt, detergent, and bile stresses should be taken into account. Although the exact survival mechanisms of bile tolerance are still unknown, lactobacillus has been shown to include many bile-related genes and components. As a biological surfactant, bile salt weakens the cell membranes of bacteria, causing leakage and death. L. plantarum was found to be resistant to a 0.3% bile concentration. Lipids are essential for maintaining the structure of cell membranes, including the fatty acid component of LAB, which is critical for bile tolerance. Bile acids frequently contain cholesterol metabolites and have the potential to damage. The damage caused by bile salts can be effectively reduced by using a combination of wild type and *bsh* gene mutation in the bile brine solution and bile tolerance of probiotics. Both the BSH activity and the rate at which bile salts were growing were decreased in *L. amylovorus*. However, because of mutational *bsh* in *Listeria monocytogenes* and *L. plantarum*, cells are susceptible to bile effects. The conjugated bile salts (CBA) that result from the conjugation of bile acids with AA or amino acids can be used to solubilize and emulsify lipids. CBAs exhibit their antimicrobial activity by disrupting homeostasis and the cell membrane. LABs, however, have unique defense mechanisms that prevent these damaging actions. The bile salt hydrolase can conjugate bile salts, which could enhance bile tolerance and bacteria survival in the gut. The bile salt hydrolase reaction can also release bile acid, creating micellars on the Bifidobacterium (Figure 3.4) membrane when bile pressure (excess) is applied(Łozińska et al., 2024).

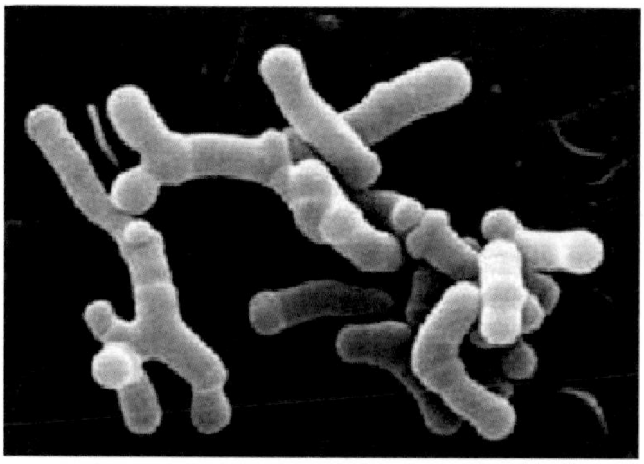

Figure 3.4 Bifidobacterium

Source: https://upload.wikimedia.org/wikipedia/commons/4/41/Bifidobacterium_longum_en_microscopie_%C3%A9lectronique.jpg,*Creative Commons License.*

For LAB to have its cholesterol-lowering effect, bile salt hydrolase, or BSH, is essential. It can link and hydrolyze conjugated bile salts to amino acids. *L. casei* may be vulnerable to bile salt stress if the *bsh* gene is absent. Many studies have indicated that probiotics have evolved BSH to deal with bile salt stress. BSH is an essential enzyme that breaks down conjugated or deconjugated bile salts into free amino acids and removes cholesterol. L. *buchneri* possesses hydrolase activity that metabolizes taurodeoxycholic acid instead of taurocholic acid. There is no correlation between BHS activity and LAB resistance to the toxicity of conjugated bile salts. The bile salts can produce protons and cause toxicity at the cell-cell interfaces, although BSH-positive cells may be able to defend themselves. Acidification can be prevented by the process of recycling and exporting the protons. Because BSH is so specific to certain forms of bile, its presence for an extended period can ensure that bacteria survive in a changing bile environment. High cholesterol is a major threat to human health. Nonetheless, probiotic-based oral medication can effectively lower blood cholesterol, and yield is unaffected by BSH activity. In rats, the human isolate L. plantarum CK102 was able to lower blood cholesterol, triglyceride, LDL-cholesterol, and free cholesterol levels. Amino acids are created by the extraction of bile salts and are then utilized as sources of carbon and nitrogen. Glycine, for example, could hydrolyze producing carbon dioxide and ammonia. But taurine hydrolyzes to produce sulfate, carbon dioxide, and ammonia.

Water moving alongside the osmotic gradient in the cytoplasm is directly impacted by changes in water activity in the extracellular space. The bacterial cytomembrane is demonstrated to exhibit considerable water permeability; yet, for the vast majority of solutes, it functions as an efficient barrier. Until the osmosis pressures on both sides of the semipermeable membrane approach equilibrium, water could enter and exit the cell. When a cell is hypotonic, the flow of water may force it to inflate and burst, or when it is hypotonic, it may lead it to shrink, dry up, and undergo plasmolysis. Aquaporins, which are water-selective channels embedded in the membrane, speed up water transport. Aquaporins are members of the major intrinsic protein or MIP family of transporters(Forrest & Bhave, 2007). The family also includes aquaglyceroporins and glycerol facilitators, which allow the transfer of many small molecules, including urea, ammonium, dihydroxyacetone, glycerol, and other polyols. By controlling the osmotic activity of the solutes in the cytoplasm, bacterial swelling can be controlled and the water content can be changed by osmosis. Osmotic regulation cells, the osmotic active substances of the Lactobacillus family, help control cell proliferation in addition to providing pressure-free protection for biomolecules in vitro. By building up glycine betaine, carnitine, and proline, many bacteria can stand against osmotic stress. Glycine betaine can shield L. plantarum from drying processes. Glycine betaine increased cell proliferation when L. plantarum was grown in a medium containing 6% NaCl. Transcriptomics demonstrated a considerable increase in the gene expression of carbohydrate transport and metabolism in the presence of glycine betaine. This could be the L. plantarum ST-III defense mechanism against salt stress. When making the dried LAB starters, sugar is known to be the primary protective agent. The equilibrium between internal and external glucose concentrations influences several of the responses of LAB to sugar and permeability. These substances are found inside the cell and come into contact with the cytoplasmic protein on both sides of the membrane.

Trehalose or Lactobacillus oligosaccharides were found to accumulate. Except for controlling intracellular solutes under hypertonic conditions, LAB quickly alters the expression of several genes that are involved in osmotic agent uptake. DnaK, GroEL, and GroES were responsible for lactate response to salt stress, according to transcriptional and proteomic research. Hyperosmotic conditions also promoted the transcription of the DNAK gene in *Enterococcus faecalis*(Lenz et al., 2010). Similar findings were observed in high salinity conditions in acidophilic Siphonococcus and a type of Lactobacillus. Another crucial component of osmoadaptation is the fluctuation in bacterial cell wall compositions with external osmolality. Bacterial membrane lipid composition is impacted by high osmotic pressure. Through lipid-protein interactions, these modifications impact the osmotic activation curves of osmotic protective agent transporters that function in *Lactobacillus lactis*. Changes could potentially impact the membrane permeability. The primary cause of the rise in cyclopropane is the alteration of the membrane fatty acid composition by high-permeable Streptococcus. In *L. bulgaricus*, the rise in sugar concentration over the entire lipid pool was similarly linked to a change in membrane function.

The rise in the total lipid sugar concentration in *Lactobacillus bulgaricus* was also linked to modifications in the membrane characteristics. There were no appreciable variations in the ratio of glycolipids to phospholipids as a result of *Lactobacillus casei* response to hyperosmotic conditions. LAB can alter the characteristics of its cell walls in response to osmotic pressure.

3.13 STRESS

Oxidative stress is due to an imbalance between the creation of reactive oxygen species and antioxidative processes. Biomolecules like lipids, proteins, and nucleic acids are damaged by ROS and oxygen, which impair their biological functioning. LABs are frequently employed as food fermentation starter cultures, so obtaining live cells is essential. Because LAB often acts as a probiotic, exposure to oxygen and reactive oxygen species during the production process has a positive impact on human health. LABs experience oxidative stress in the gastrointestinal tract due to oxygen gradients and the immune system, which lowers the number of viable cells. Oxidative stressors had a major impact on LAB gene expression, growth, and survival. Stress reset the LAB gene expression, and to survive, the physiology changed accordingly. Glyceraldehyde-3-phosphate dehydrogenase catalyzes the conversion of glyceraldehyde-3-phosphate to glycerate-1,3-biphosphate during glycolysis in *Lactococcus lactis*(Cesselin et al., 2011). The *gapB* gene is highly expressed via the glycolytic pathway. The *gapB* gene has been explored in 2D gels. When its cells are exposed to O_2, it manifests as two spots with distinct isoelectric points but the same molecular weight. It was shown that under these conditions, one in a thioredoxin reductase mutant (trxB1) and the other in respiration metabolism, the relative levels of these two *Lactococcus lactis* spots had changed. Nevertheless, oxidative stress is connected to both of these conditions. Before the GapB protein is attacked, the thioredoxin–thioredoxin reductase system must eliminate reactive oxygen species (ROS), such as O_2, superoxide, H_2O_2, and hydroxyl radical(Allakhverdiev, 2015). *Lactococcus lactis* GapB can prevent the flux of glycolysis from decreasing due to GapB oxidation while maintaining full glycolysis activity. In the presence of glutathione, cysteine, and pyruvate, the growth of lacking texB1 slowed down. Pyruvate formate lyase (PFL) facilitated the conversion of pyruvate to acetyl-CoA during anaerobiosis. When PFL transitions from anaerobiosis to aerobic respiration, it splits into two pieces, which ultimately cause its permanent inactivation. This cleavage might be prevented by PFL decativase converting the glycyl radical into glycine. An alternate metabolic pathway was activated to maintain the generation of acetyl-CoA. The molecule catalyzes the conversion of pyruvate to acetyl-CoA. Under conditions of aeration, flavoprotein is essential for the growth of many bacteria.

A large amount of acetate and acetoin were produced at the exchange of lactate by a NADH:H_2O of *Lactococcus lactis* that causes oxidase to overproduce, suggesting that the enzyme can efficiently increase the conversion of pyruvate to acetate.

Studies demonstrate that heme-dependent cytochrome oxidase can be activated by *Lactococcus lactis*, *E. faecalis*, and other LAB, establishing a complete respiration chain. Heme-dependent cytochrome oxidase CydAB is often believed to have a higher affinity for O_2. ROS production is inhibited by the membrane consumption of O_2, which lowers the amount of O_2 in the cytoplasm. *Lactococcus lactis* and other LAB manage their heme repository due to ROS reactivity. However, allowing sufficient heme allows for the activation of the CydAB cytochrome oxidase. Two TCSs in *Lactococcus lactis* were shown to be involved in oxidative stress: LlrF (DNA-binding regulator) and LlkinF (sensor) (Mailloux, 2020). Only 9% of mutants that have had LlrF inactivated survive when exposed to 4 Mm H_2O_2. LlrF inactivation is particularly susceptible to peroxide. It appears that a TCS in *Streptococcus mutans* termed ScnRK may control several stress-related genes, such as TPX, which may encode the enzyme mercaptan peroxidase. The PerR gene and the OhrR gene of *Bacillus subtilis*, which are linked to the peroxide stress response, were found in the genome of *Lactococcus lactis*. The iron-containing repressor family is linked to the PerR gene in Bacillus subtilis. The $HO^{\cdot}$ oxidized histidine at positions 37 or 91 in the peptide sequence, and the PerR reacted with peroxide via iron through the Fenton reaction. Studies revealed that an oxidative stress response could be triggered by a regulator that is comparable to PerR in *E. faecalis*. The homogeneous dimer OHRR protein in Bacillus subtilis is a member of the MarR (antibiotic resistance) families. The thior peroxidase-encoding genes may be regulated by the OHRR protein, leading to the reduction of hydrogen peroxide to alcohol. Because its distinct cysteine residue has not been oxidized, OHRR is immune to hydrogen peroxide stress at the site of peptides at position 15. Under oxidative stress, cysteine is oxidized to sulfenic acid or to a higher oxidation state.

Chapter 4

RHIZOSPHERE AND MICROBES

Contents

4.1 INTRODUCTION

The small area of soil surrounding a plant root is known as the rhizosphere, and it is impacted by a number of elements, including the soil microorganisms that live there and the root exudates. Because it gets nutrients from the soil and from the microorganisms surrounding the root, the rhizosphere is thought to be the most active part of the soil. Many rhizosphere effects, such as exudation of some organic compounds or specific ionic absorption, could be the origin of many of these effects. Such rhizosphere (Figure 4.1) impacts have been repeatedly studied for their ecological significance. The volume of soil next to and affected by a plant root is known as the rhizosphere, and it is thought to be a hot spot for microbial colonization and activity.

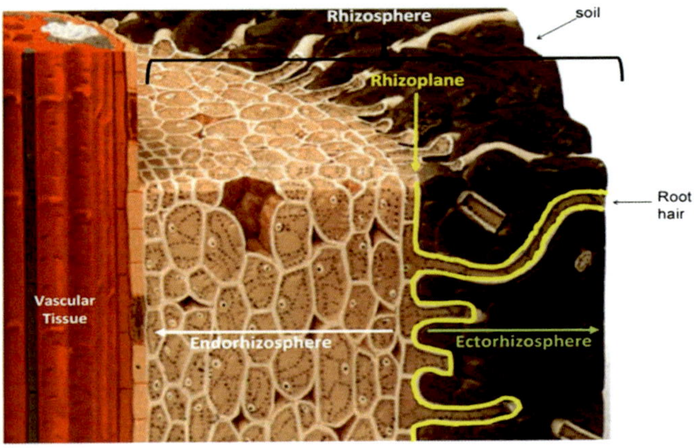

Figure 4.1 Rhizosphere

Source: https://upload.wikimedia.org/wikipedia/commons/2/27/Rhizosphere_composition.jpg,*Creative Commons License.*

Rhizospheres get high concentrations of nutrients produced by plant photosynthesis as opposed to bulk soil, which only contains minimal amounts of accessible organic carbon sources. In soil-plant systems, microorganisms in the rhizosphere of plants control the cycling of nutrients. By either extending the root system (fungal hyphae) or solubilizing macronutrients like phosphorus or sulfur, rhizosphere microorganisms improve plants' capacity to absorb nutrients from soil. Mutualistic relationships between soil microorganisms like arbuscular mycorrhizal (AM) fungi (Figure 4.2) and plant growth in nutrient-poor situations have been found(Linderman, 2015). Plant-pathogen interactions are also influenced by microorganisms in the soil. Bacteria that have been injected/inoculated into cucumber (*Cucumis sativus*) roots can reach the root and cause systemic resistance against a variety of cucumber diseases.

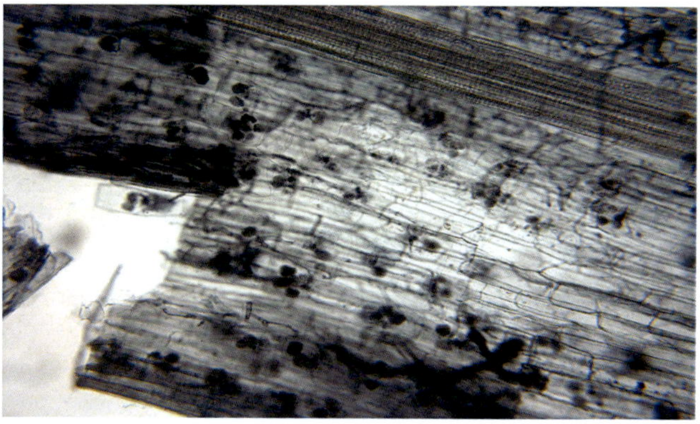

Figure 4.2 Arbuscular mycorrhizal (AM) fungi

Source: https://upload.wikimedia.org/wikipedia/commons/8/86/Arbuscular_mycorrhiza_microscope.jpg,*Creative Commons License.*

Many environmental factors such as soil type, nutrient status, pH, texture, organic matter content, moisture, interactions of these variables, etc. as well as plant factors affect the abundance and activities of soil microorganisms. The ability of various fluorescent pseudomonad strains introduced to survive varies depending on the texture of the soil. However, as soil microbial growth is carbon-constrained/limited, organic matter content has a huge impact on microbial populations. It has been demonstrated that the release of exudates and lysates in the rhizosphere specifically stimulates species of Pseudomonas, Flavobacterium, Alcaligenes, and Agrobacterium. It has been suggested that one of the main factors influencing the diversity of microbes in the rhizospheres of various plant species is the range of organic chemicals emitted by plants.

4.2 RHIZOSPHERE ALTERATIONS

The rhizosphere is intricate in terms of chemistry, physics, and biology. Approximately 20% of the assimilation of photosynthetic chemicals by plant roots is released as water-soluble sugars, amino acids, vitamins, and hormones; root exudates also contain organic acids, amino compounds, and sugar-phosphate ester. It has been hypothesized that the low molecular weight exudates, lysates, gases (CO2 and ethylene), mucus, and organic compounds that are released are important factors that affect the diversity and development rates of microorganisms in the rhizosphere of various plant species. The impact of root exudates is contingent upon their ability to disperse away from the rhizoplane. Root exudates are important to identify and measure because they change the way nutrients travel through the rhizosphere and promote organic soil changes. The size and charge of the compound as well as the characteristics of the soil, such as its texture, pH, moisture content, and temperature, all affect how quickly a compound diffuses into soil. Water-soluble compounds are the easily cleaved organic molecules released by roots; these compounds provide nutrition to the microbial populations that colonize the surfaces of the roots. The microorganism can alter the soil root environment or produce different microbial metabolites that impact plant physiology and growth. The root exudates have the potential to modify the composition of microorganisms that colonize the rhizosphere by creating a niche that determines which bacteria will grow. Because the majority of microorganisms are oligotrophic and react slowly to root exudation, the impact of root exudates on dominating bacterial species may be negligible. The addition of low molecular substrates that are passively supplied by roots favors copiotrophs.

4.3 METABOLITE ANALYSIS

Understanding the concentration and composition/ingredients of organic materials exuded by the roots of various plant species provides insight into the biological and chemical processes taking place in the rhizosphere. Primary metabolites include sugars, amino acids, sugar alcohols, and organic acids, and secondary metabolites are included in root exudates. Secondary metabolites are primarily responsible for the creation of microbial biodiversity. Many terpenoids, coumarins, and flavonoids are also accumulated and released by plant roots.

Many environmental parameters, such as pH, soil type, oxygen status, light intensity, soil temperature, nutrient availability, and the presence of microbes, have an impact on the composition of root exudates. These variables might affect root exudation more than species-specific variations. Because soil texture changes the mechanical impedance to the root, it has been demonstrated to have an impact on root exudates.

The amount of clay also affects how various peptides and amino acids bond to one another. Root exudate quantities and composition are stimulated by microorganisms(G. Singh & Mukerji, 2006). The age and species of a plant are known to affect the root exudate composition. Plant species can exhibit varying patterns of root exudation, and soil microorganisms have been demonstrated to react to plant exudation. The amount of amino acids released by peas and oats varies. Different pines have distinct patterns of exudation. Mature trees are known to exude more varied and abundant amino acids, but young maples are known to ooze more carbohydrates. Researchers discovered a relationship between the microbial biomass C under different annual and perennial species and the water-soluble organic C, indicating variations in C deposition from the plant species. Clover had a higher exudation rate than wheat, which was higher than ryegrass. Legumes also have a lower C-to-N ratio (<20) than grasses (> 30). This proves that legumes release higher amounts of amino nitrogen into the rhizosphere than wheat does, which may lead to the emergence of distinct microorganisms there.

Figure 4.3 Corn

Source: https://upload.wikimedia.org/wikipedia/commons/7/79/VegCorn.jpg,*Creative Commons License.*

4.4 ORGANIC ACIDS AND EXUDATES

Important crop rhizospheres may contain organic acids other than tartaric. In Zea mays (corn), (Figure 4.3) citric acid was found to predominate and tartaric acid was present only in trace amounts. Citric acid and succinic acid were the predominant organic acids in root exudate solutions of the same plant species(Zhang et al., 2014). It was discovered that acetic and oxalic acids predominated in wheat plants (*Triticum turgidum var. durum L.*). The exudate from the roots of flax (*Linum usitatissimum L.*) was discovered to be mostly acetate. Enhanced rates of exudation of organic acids by plants are in response to toxic levels of aluminum in soil or enhanced accumulation of less soluble plant nutrients, such as P, K, and Fe, even though the mechanism by which plants control root exudates is not fully understood. P-deficiency and aluminum toxicity are linked to increased exudation of citrate and malate by plant roots. Zinc deficiency (soils) causes higher rates of root exudates. The availability of inorganic plant nutrients including Mn, P, and Fe is influenced by root exudates. Microorganisms in the rooting media have the potential to either accelerate the breakdown of organic root exudates or stimulate the exudation of organic materials by plant roots. *Pseudomonas fluorescens* and other microorganisms can release organic molecules called siderophores into the rhizosphere. This can have an impact on the uptake and availability of specific mineral nutrients, particularly iron. Bacteria are drawn to roots by the chemoattractant properties of root exudates. Sugars or sugar alcohols may serve as signals to soil fungi and determine fungal diversity in natural rhizosphere habitats. The growth of endophytic fungus was promoted by the sugar alcohol mannitol. External sugar concentrations may have an impact on the expression of genes in ectomycorrhizal basidiomycetes, even while non-host carbon sources were unable to stimulate the fungi. In addition to serving as transcriptional signals during nodule formation, BNF production by plants, and interactions between plants and arbuscular mycorrhizae, flavonoids found in root exudates also function as a means of communication between soil bacteria and host plants. These secondary metabolites function as a signaling agent for the microflora as well as a defense strategy for plants against diseases. Legumes release different types of flavonoids, such as chalcones, which cause *Agrobacterium tumefaciens* to produce virgenes and Rhizobium and Bradyrhizobium to express nodulation (nod) genes.

4.5 HYPHAL DEVELOPMENT

Both in vitro and during the pre-colonization stage of AMF production, plant/host factors promote the growth of AMF hyphae. Plants with arbuscular mycorrhizal associations have increased root permeability and/or exudation. Cultures of Ri-T-DNA-transformed roots and stressed suspension cultures of legume cells demonstrate the quality of root exudates, which promotes AMF colonization. Several substances found in root exudates function as signal molecules that can stimulate or induce hyphal development, branching differentiation, and host penetration. Phenolics, which are known to function as signal molecules in plant-microbe interactions, are among the several potential hydrophobic chemicals found in root exudates. Phosphate deficit caused an increase in isoflavone in clover leaves, while parsley with low phosphate levels showed the presence of some unidentified chemicals. Two isoflavonoids that function as signal molecules in AM-symbiosis were identified from root exudates of clover roots grown under phosphate stress: formononetin (7-hydroxy, 4' methoxy isoflavone) and biochanin A (5,7-dihydroxy, 4' methoxy isoflavone)(Heinonen et al., 2002). It has been discovered that hesperetin, naringenin, and apigenin promote G. margarita hyphal development. Flavones, flavanons, and isoflavones were inhibitory, only flavonoids stimulated fungal growth. The hyphal growth from germinating spores was enhanced by organic acids found in root exudates, while the hyphal growth from *G. caledonicus* spores was increased by amino acids such as cystine, glycine, and thiamine.

Figure 4.4 *Gigaspora margarita*

Source: https://upload.wikimedia.org/wikipedia/commons/c/c7/Gigaspora_margarita.JPG,*Creative Commons License.*

4.6 EXUDATE DYNAMICS

The amount and quality of root exudates are influenced by the phosphorus nutrition of the host plants. Phosphorus affects the root colonization of arbuscular mycorrhizal fungi in the soil solution and in the tissues of the host plant. Plants with a P deficit exude more organic acids, sugars, and amino acids from their roots. It has been demonstrated that the root exudates of P-deficient citrus, clover, onion, and carrot stimulate the growth of arbuscular mycorrhiza (hyphae). *Gigaspora margarita*(Figure 4.4) has been shown to generate secondary appressoria on the onion roots. *Pueraria phaseodes* suspension-cultured cell exudates promoted the growth of arbuscular mycorrhizal hyphae. Compared to root exudates produced in the presence of P, Ri-T-DNA-transformed *Daucus carota* grown in the absence of P produced excess hyphal development from G. margarita spores. The hyphal growth of *G. margarita* was similarly impacted by root exudates from onions with distinctly varied P status. Phosphate treatment to shoots prevented the production of secondary appressoria, showing the inhibitory influence of plant P status. These findings demonstrate that the host plant's phosphorus nutrition affects root exudates, which in turn affects the degree of root colonization by arbuscular mycorrhizal fungi. Mycorrhizal infection had no discernible impact on the quantity or composition of root exudate in maize at optimal P levels.

Arbuscular mycorrhizal fungi may be impacted by variables other than P in root exudates. Root exudates of carrot and clover encouraged the hyphal growth of *G. margarita*. Hyphal morphogenesis, which promotes root contact and appressorium production, can be elicited by host root exudates. In the development of appressorium on isolated host cell walls, a chemical signal is not necessary for contact recognition. The plant species has an impact on the introduced bioinoculant population dynamics. The ability of the rhizosphere communities of clover, ryegrass, and wheat to use a range of substrates varies. The variations in the type and amount of carbon sources produced in the root exudates primarily control the composition of microbial communities specific to each plant in their rhizospheres. Transgenic cultivar rhizosphere bacterial composition allowed for differentiation from non-engineered plants. Genetic engineering did not, however, substantially alter the composition of the microbial groups in the rhizospheres of potatoes. In the rhizosphere of several plant species, such as wheat, clover and ryegrass, and grasses, there exist varied microbial populations. In the rhizospheres of wheat, maize, and grass, actinomycetes made up a significant portion of the microbial population. Pseudomonads, on the other hand, were only slightly stimulated in these environments. Some researchers have also documented the selective activation of Pseudomonas species in the rhizosphere(Song et al., 2021). Certain genetically distinct species of *Burkholderia cepacia* are present in maize crop cultivars. Researchers separated antagonistic rhizobacteria with varying genotypes and phenotypes from various Verticillum host plants. In comparison to the control soil, they found that the number of isolates with antagonistic activity was maximum for the rhizosphere of strawberries, followed by oilseed rape and potatoes. The quantity and composition of the antagonists against Verticillium varied according to the plant species.

4.7 PROFILING OF BACTERIAL COMMUNITIES

Every plant species chooses a particular bacterial community from the groups of bacteria found on seeds or in soil. Research using rRNA-gene profiling methods and community-level physiological profiles show that plant species, rather than other variables like soil origin or agricultural treatments, play a major role in the selection of bacterial communities in rhizospheres. In the rhizospheres of crops grown in soil, plant age-dependent effects (bacterial community) are also noted. Different root exudate compositions during various growth stages result in this change. Plant genotype also affects the rhizosphere community composition, albeit there are some outliers. The 16S RNA gene of *Rhizobium leguminosarum*(Figure 4.5) matched the predicted profile of the dominant operational taxonomic unit (OTU) in the rhizosphere of *Trifolium repens* (Figure 4.6) (Hartman et al., 2017).

Figure 4.5 *Rhizobium leguminosarum*

Source: https://upload.wikimedia.org/wikipedia/commons/6/60/Soybean-root-nodules.jpg,*Creative Commons License.*

Figure 4.6 *Trifolium repens*

Source: https://upload.wikimedia.org/wikipedia/commons/5/52/Trifolium_repens_-_white_clover_on_way_from_Govindghat_to_Gangria_at_Valley_of_Flowers_National_Park_-_during_LGFC_-_VOF_2019_%281%29.jpg,*Creative Commons License.*

In bulk soil, this OTU was discovered at a significantly lower level. Plant roots are responsible for the selection of bacterial populations linked to plant roots. The capacities of populations derived from the roots of each plant species as well as from the uncultivated soil to utilize various organic compounds and assimilate nitrogen varied. The morphological and genotypic characteristics of fluorescent pseudomonad populations isolated from uncultivated soil and the roots of two plant species, flax, and tomato, were different. The quality and amount of root exudates generated by the various wheat genotypes are influenced by the soil micronutrient status, resulting in differences in the degree of bacterial colonization of roots between the genotypes. The soil microflora easily absorbs exudates from roots. Higher bacterial populations in the rhizosphere and higher exudation from wheat roots were the results of increased N fertilizer treatment. When taxonomically analyzing plant endophytic microbial communities, the selection pressure becomes more evident. Variations in root morphology or root exudates could be the cause of the observed variances in the endophytic population between the cultivars. A close link between the two communities is suggested by the fact that endophytic bacterial taxa are a subset of rhizosphere bacterial taxa. Bacteria can enter the interior of roots by either hydrolyzing cellulose or by penetrating tumor lesions caused by auxin at the lateral root branching.

Figure 4.7 *Arabidopsis thaliana*

Source: https://upload.wikimedia.org/wikipedia/commons/6/6f/Arabidopsis_thaliana.jpg,*Creative Commons License.*

The genotypic and phenotypic diversity of a PhlD-containing Pseudomonas strain isolated from the rhizosphere of wheat crops was investigated. The genotype of *Arabidopsis thaliana*(Figure 4.7) has a considerable impact on the genotypic structure of *Pseudomonas brassicacearum* populations as determined by REP-PCR fingerprints(Berg et al., 2002). There are variations in the patterns of root exudates and rhizodeposition within and across plant species. The kinds and quantities of root exudates and rhizodeposition are correlated with the activities of rhizosphere bacteria. Higher levels of carboxylic acids, such as succinic acid, malic acid, and oxalic acid, were exuded by the roots of wheat cultivars, which varied in their associative N2 fixation by *Azospirillum brasilence*. The amounts and types of Fe-producing and chelating chemicals released by tomato mutants with varying iron-mobilization characteristics were observed in their roots. The quantity of amino acids and oxalic acid they released varied.

Figure 4.8 Chickpeas

Source: https://upload.wikimedia.org/wikipedia/commons/e/ea/Sa-whitegreen-chickpea.jpg,*Creative Commons License.*

Chickpeas (Figure 4.8) produce a lot of organic acids, especially when the pH is low. This may cause microbes that are tolerant of low pH to thrive on the substances and the byproducts of their breakdown. Crops with better traits are developed through genetic manipulation.

The release of gene products or changes in the composition of root exudates from genetically modified plants may cause qualitative and quantitative changes in the bacterial consortiums and soil environment in the rhizosphere.

The herbicidal chemicals glufosinate, a L amino acid that inhibits glutamine synthetase, are ineffective against transgenic crops such as maize, rape, and sugarbeet. A few studies did, however, also show that the soil microflora was sensitive to this herbicide, which could lead to a reduction in soil fertility. There are transgenics for other traits as well. Researchers observed changes or disruptions in the rhizosphere community structure of Eubacterium and Pseudomonas due to the introduction of transgenic *Brassica napus* that is glufosinate-tolerant.

The rhizosphere of transgenic plants had changed microbial communities as a result of changed root exudation, according to DGGE analysis of 16S rRNA gene fragments amplified from rhizospheric eubacterial and Pseudomonas communities.

4.8 DIVERSITY

Some researchers demonstrated that plant species and cultivars had a substantial impact on the diversity and composition of rhizosphere bacterial populations in agricultural soils. Fewer Arthrobacter and Bacillus isolates were found in the rhizosphere of a transgenic Canola cultivar than in the rhizosphere of an untransformed Canola cultivar. The observed variations could have been caused by variations in the pattern of root exudation. Bacillus predominated in the rhizosphere of wheat but not in the canola rhizosphere, indicating differences in the rhizosphere bacterial communities. The amount and composition of root organic acid generated and released into the rhizosphere under field settings varied between transgenic alfalfa cultivars over-expressing a nodule-enhanced malate dehydrogenase (ne MDH) and the untransformed alfalfa cultivar. The nucleotide sequencing of PCR-based 16S ribosomal DNA (r DNA) and biological microtiter plate analysis of rhizospheric microbial populations revealed qualitative differences in the abundance of bacterial groups between rhizosphere soils of transgenic and untransformed alfalfa. Compared to non-transformed alfalfa, biology investigations revealed a higher degree of microbial functional diversity in the transgenic plant rhizosphere. The distribution and relative abundance of the various phylogenetic groups changed in response to a transgenic alfalfa line that produced 7.1 times more organic acids than untransformed alfalfa. This led to qualitative differences in the bacterial community structure between the rhizospheres of the transgenic and untransformed alfalfa. It has been demonstrated that Pseudomonas sp. colonizes roots via the use of organic acids in root exudates. Because the rhizosphere of various plant species supports the added or introduced bioinoculant unequally, the degree of root colonization could be improved by matching rhizobacteria genotypes with crops. The potential function of root exudates in the interaction between plants and microbes, the mobilization of nutrients, and the adoption of crops in stimulating soil conditions has applications in the engineering of plants suitable for a range of agricultural systems. Transgenic plants that have a different root exudate composition from wild-type plants can be useful tools for determining the ecological importance of different root exudate components in microbial ecology. The root, the soil that is linked to the root, and the nearby soil that the root influences are all part of the rhizosphere, which is the rooting zone of plants. It is a densely populated area where the roots must compete for space, water, and mineral nutrients with the invading root systems of nearby plant species. The soil is home to a variety of microorganisms, such as bacteria, fungi, and insects, which feed on an abundance of organic materials. The interactions between roots and microbes, as well as between roots and insects, are probably ongoing in this biologically active soil zone. Communication between roots and microbes can be either beneficial (symbiotic) to the plant, as in the case of nitrogen-fixing bacteria, mycorrhizal fungi, and epiphytes, or detrimental as in the case of interactions with parasitic plants and pathogens, such as bacteria, fungi, insects, etc.

4.9 ENDOPHYTIC BACTERIA

The endophytic bacteria that are able to colonize root tissue are now included in the term rhizosphere(Cocking, 2003). This model creates a continuous apoplastic channel from the root epidermis to the shoot, which is sufficient for the migration of microorganisms into the xylem because the root cortex becomes a part of the soil and root microbial habitat. As a result, a range of microorganisms linked to roots are produced, capable of living in the rhizosphere, the root cortex, and other plant organs. Given the intricacy of the underground environment, roots are not the passive targets of soil organisms. Apart from offering structural stability and permitting the assimilation of water and nutrients, roots also carry out specific functions such as the production, accumulation, and release of a wide range of substances. Through root exudates, between 5 and 23 percent of all photosynthetically fixed carbon is transported to the rhizosphere. In the rhizosphere, the substances released by plant roots play crucial functions as chemical attractants and repellents. Plants impact the diversity and composition of microorganisms in the rhizosphere soil by exuding a wide range of soluble and readily diffusible substances such as proteins, lignin, and cellulose. The rhizosphere microbial connections are dynamic by nature. The efficacy of these microbial partnerships may be influenced by the duration of root survival, which suggests that long-lived perennials might be most advantageous. The specific rhizobial associations formed with legume roots must be regenerated with freshly emerging roots to continue for the duration that the root is alive.

One important aspect influencing how soil organisms colonize the rhizosphere is the type and quality of the carbon supply that flows from the root. Organisms must either migrate forward with the root tip, towards the root tip, or in some other means obstruct the root tip's growth in order to reach it. The sheath of up to 10,000 live border cells that are created daily by the developing root cap and discharged into the surrounding rhizosphere is a potentially vital component of this system. It was proposed that these border cells serve as protective walls for pathogenic organisms in the soil surrounding the root tips, lowering the likelihood of a pathogenic attack at the root tip. The border cells at the root tip may separate before the diseased root tip itself if the pathogen infection does take place there. Likewise, the molecules that roots release can be altered by soil microbes, which allows them to create a favorable environment. Living near roots, bacteria and fungi consume dead root cells and root exudates. There is competition among the microbiological species here. Bacteria employ a variety of strategies in their struggle for carbon compounds, or organic matter, including the synthesis of antibiotics and toxic substances that eliminate competitors. Bacteria produce compounds that encourage a plant to grow more roots, expand the area of its roots to allow for colonization, and produce more root exudates. The interactions between roots and microorganisms in the rhizosphere are influenced by the entire biological population, not just the microflora found immediately beneath the roots. These interactions are governed by intricate molecular signaling. Plants and soil microorganisms have co-evolved, leading to microbial reactions to plant exudation and plant exudation reactions to microbe presence(Lyu et al., 2021).

Figure 4.9 Sweet basil

Source: https://upload.wikimedia.org/wikipedia/commons/9/90/Basil-Basilico-Ocimum_basilicum-albahaca.jpg,*Creative Commons License.*

As a plant eventually engages in positive or negative communications, depending on the other elements in the rhizosphere, root secretions play symbiotic or defensive roles. The flavonoids found in legume root exudates stimulate the *Rhizobium meliloti* genes in the nodulation process. On the other hand, phytoalexins, defense proteins, and other unidentified compounds secreted by pathogenic bacteria wage a constant underground chemical attack to preserve the vulnerable and physically exposed root cells from their constant attacks.

4.10 FUNGAL EXTRACTS

In the root exudates of sweet basil (*Ocimum basilicum*), (Figure 4.9) rosmarinic acid (RA) was found, induced by fungal cell wall extracts from *Phytopohthora cinnamomi*(Kapoor & Mukerji, 2006). In P-deficient Lupinus albus, root exudation has been linked to the dynamics of rhizosphere microbial populations. The alternating root exudation during cluster root development has an impact on the bacterial population on the roots, which is what causes the significant variations in the rhizosphere soil composition. The amounts of citrate, malate, and cis-aconitate in the rhizosphere soil solution, but not fumarate, had an impact on the bacterial population on cluster roots. Citrate exudation affected fungal populations in cluster roots but not in other root zones. The white lupin cluster roots are functionally associated with an effective chemical mobilization of soil P-sources that are sparingly soluble. This process is facilitated by ectoenzymes (acid phosphatases) and organic chelators (citrate, malate, and phenolics), which release large amounts of these P-sources into the rhizosphere. Significant changes in the rhizosphere pH, redox potential, carboxylate and phenolic acid release, and high chitinase expression indicate a substantial impact on the microbial communities during the growth of cluster roots. The microbial turnover of root exudates and the synthesis of microbial metabolites involved in nutrient mobilization may be impacted by this. The number of microbes is also influenced by the age of the plant. Small diffusible signaling molecules, such as peptide-signaling molecules for Gram-positive bacteria and acylated homo-lactones (AHLs) for Gram-negative bacteria, promote cell-to-cell communication between bacteria. An auto-inducer triggers the activation of transcriptional activator proteins, which in turn induce specific genes when it reaches a threshold concentration at high population densities. Therefore, a bacterial population can regulate gene expression in response to cell density (quorum sensing) through intercellular communication. It was observed that several plant crude aqueous extracts showed AHL inhibitory action. It was discovered that the root exudates of Pisum sativum (pea) seedlings included several bioactive chemicals that, when present in bacterial reporter strains, replicated AHL signals and stimulated AHL-regulated behaviors in other strains. Therefore, roots may have evolved defense mechanisms by secreting substances like signal mimics, signal blockers, and/or signal-degrading enzymes into the rhizosphere that obstruct bacterial quorum sensing responses. The chemical composition of these active mimic secondary metabolites remains partly determined.

The largest repository of biological diversity is found in soil microbes, which are also essential to the health of terrestrial ecosystems. Particularly at the size that microorganisms inhabit, soil offers a temporally and geographically varied environment. Heterotrophic microbial communities that live in soil mediate important processes that regulate the cycling of carbon (C) and nitrogen (N) in ecosystems and may also serve as a link between plant variety/diversity and ecosystem function. Microbial communities and the ecosystem process they mediate are predicted to be impacted by plant diversity. The availability of growth-limiting resources determines the composition of biotic communities. Organic compounds found in dead leaves and roots, or detritus, limit the amount of resources available to soil microbial communities because they can be used to produce cellular energy. Because the biochemical composition of plant species varies, shifts in plant diversity may also affect the production of organic compounds in detritus, which in turn affects the composition and activity of heterotrophic microbial communities. Plant diversity greatly increases the rates of microbial processes that mediate the cycling of carbon and nitrogen in ecosystems; the effect is more reliant on plant production than species richness. Microbial respiration releases photosynthetically fixed carbon back into the atmosphere during the breakdown of organic materials. More soil water is used by plants in species-rich communities than by those in less diverse groups. As plant diversity increases, this response would lower soil matric potential, favoring fungal dominance and changing the function of microbial communities.

The rise in microbial respiration is consistent with the pattern of N mineralization. Plant diversity affects the composition and function of microbial communities, which in turn increases the supply of soil N to plants and contributes to greater productivity in the majority of species with experimental plant communities, even though the precise mechanisms causing greater rates of mineralization are still unidentified. These correlations might not apply to all microbial species. It was discovered that the genetic diversity of ammonia-oxidizing bacteria (the initial stage of nitrification) did not react to variations in the diversity of the above-ground vegetation in grasslands. Microbial populations on the root surface are impacted by plants, not the soil in the rhizosphere(Hawkes et al., 2007). The dominating microbial communities appear to be extremely stable in both space and time outside of this zone. Plants seem to regularly choose a unique microbial community on the root surface, which is made up of a wide variety of taxonomic classes of microbes. Wheat, ryegrass, and clover were shown to have a selective influence on microbial diversity in the rhizosphere. The microbial population of the various plant rhizospheres was differentiated between their carbon sources. The primary substrates that cause this differentiation are carbohydrates, carboxylic acids, and amino acids, indicating that plants may differ in how these molecules are exuded. The often-found bacterial genera in the rhizosphere include non-culturable Verrucomicrobium and Acidobacterium, as well as several well-known species like Bacillus, Pseudomonas, Stenotrophomonas, and Comamonas. The composition of the microbial species in the rhizosphere is influenced by plant genotype. Wheat genotypes affect the microbial population in the rhizosphere of wheat. Compared to hexaploid wheat, the rhizosphere of wild diploid and tetraploid wheat had higher populations of bacteria and actinomycetes. The diploid DD and VV chromosomes primarily encouraged these two types of bacteria. Secondary metabolites have been shown to inhibit the growth of mycelium in a number of fungi that have been isolated from the rhizosphere. Under field conditions, the composition of the microbial community is affected differently by genotypes that differ in their synthesis of these chemicals.

4.11 LYSOZYME AND OTHER TRANSGENES

T4 lysozyme-expressing transgenic plants are effective in treating a range of diseases that affect commercially crop plants(Widmer, 2007). T4 lysozyme affects both Gram-positive and Gram-negative bacteria through a non-enzymatic mechanism that may include membrane rupture or by its muramidase activity against the bacterial cell wall component murein. Erwinia carotovora infections are less likely to affect transgenic potato plants as a result of this genetic change in the plant. A noticeable quantity of T4 lysozyme is also released from the roots, resulting in bactericidal action on the root surface. Many bacterial species related to plants exhibited susceptibility in vitro. The expression of T4 lysozyme alter the bacterial population in the rhizosphere, which could have an impact on the function or quality of the soil, given that the effect on various species differed greatly. When compared to rhizosphere communities of wild-type plants and transgenic controls lacking the lysozyme gene, the rhizospheric bacterial communities of transgenic potato lines that produce T4 lysozyme for protection against bacterial infections did not exhibit any apparent differences. The rhizospheric community was found to be influenced by ambient conditions rather than the transgenic plant T4 lysozyme expression, as indicated by DGGE profiles, bacterial isolate compositions, and Biolog profiles of several samples. The transgenic maize had pat genes that conferred resistance to the herbicide glufosinate. There was no difference between the rhizospheric microbial SSCP characteristics. Bacterial populations from the rhizosphere of sugar beet plants cultivated in field conditions have distinct PCR-SSCP patterns. Different microbial communities were linked to the roots of glyphosate-tolerant oilseed rape plants, both transgenic and non-transgenic, that were grown in fields. Therefore, in addition to the genetic insertion that confers herbicide resistance, the investigators were unable to rule out the possibility of other genotypic changes. Soils surrounding the transgenic lines included a variety of strains that were not present in the control treatments. There was more Rhizobium diversity in the fields of some transgenic lines. No appreciable effects on soil basal respiration or rhizobial populations were noted. Significant effects on microbial communities were observed as a result of accidentally altered plant traits through genetic engineering. The *Bacillus thuringiensis* endotoxin-expressing cotton microbial DNA fingerprints, total bacterial counts, and species composition were compared to those of the wild-type plant in several studies. Over the past 90 years, the quantity of CO_2 in the atmosphere has increased, and this increase is expected to continue at an average rate of 1.5 ppm per year(Arnell et al., 2002). In response to increased CO_2, plants fix more carbon, which increases plant biomass and the amount of carbon allocated below ground. Changes in the amount and quality of carbon that enters the soil through rhizodeposition or litter under conditions of increased CO_2 will have a substantial impact on the composition of the soil microbial community, which will affect nutrient cycling and acquisition. The pace of decomposition is specifically influenced by the C:N ratio of plant litter. It has been proposed that this rise in the C:N ratio may slow down the breakdown rate of litter, which will lower the amount of N that plants can access and mineralize. When plants are grown at increased CO_2, the rhizospheric microbial community alters, with fungal growth being preferentially stimulated. The gross exuding molecules with greater C:N ratios were the cause of these alterations. Microbial communities from the rhizosphere of increased CO_2 growing plants exhibited a greater utilization of all C sources, except those having a low C:N ratio like neutral and acidic amino acids, amides, N-heterocycles, and long chain aliphatics. because fungi grow on more complicated substrates and are more effective than bacteria in converting substrate into biomass, the preferential stimulation of fungus under elevated CO_2 levels may be caused by a rise in the amount but the decrease in the quality of litter.

The rhizosphere responded most strongly to increased atmospheric CO_2 in areas where the number of microarthropods seen on roots increased, arbuscular mycorrhizal root length increased, and microbial biomass remained unchanged while protozoa numbers increased, suggesting faster bacterial turnover. The primary force behind the breakdown of litter and the cycling of nutrients in soils is the microbial community. The soil system is continuously receiving easily obtainable types of nitrogen and carbon from the rhizosphere. As a result, the microbial biomass grows more rapidly. The measures of microbial biomass have been widely employed as markers of stress and disturbance in ecosystems. The values of microbial biomass varied greatly throughout the forest stand. Various constraints were observed to be influencing the expansion of the microbial biomass across various forest types. Microbial biomass and soil pH showed a substantial correlation in the more acidic soils. In soils of all forest types, positive associations between total soil N, soil moisture, and microbial biomass were discovered. Low molecular weight chemicals get into the soil by microbial metabolism, rhizodeposition, plant litter decomposition, and decomposing soil organisms. These materials can deplete quickly once they are in soil; amino acids, for example, typically have half-lives of one to twelve hours. Therefore, the concentration of sugars and amino acids depends not only on the quantity, quality, and pace of rhizodeposition and litter breakdown but also on the exudate turnover durations. High sugar and amino acid concentrations may be a sign of both low turnout rate and high substrate abundance. The local climate and agricultural methods can cause major changes in the soil environment during a growth season, which can affect the microbial populations in the soil and the dynamics of nutrients. The size of the microbial community was found to fluctuate only less, and the concentration of microorganisms in the top layer was consistently higher. The soil bacterial capacity to use carbon varied with time, and patterns of color creation indicated a drop in microbial activity.

4.12 CROPS AND ASSOCIATED MICROBES

Crop development and yield may be enhanced in specific conditions by free-living diazotrophic bacteria that bind to the roots of non-leguminous plants(Franche et al., 2009). However, it is believed that only a small percentage of the nitrogen absorbed directly by plants comes from nitrogen fixation by free-living rhizobacteria. The observed growth responses are attributed to the secretion of substances that promote plant growth. Because there are no morphological changes resembling nodulation, the relationships between plants and diazotrophic bacteria have not been regarded as symbiosis. However, it has been shown that significant amounts of fixed nitrogen can be transferred directly to the plant, especially in salt marshes where the Spartina predominates and tropical grasses. These relationships show that plants and diazotrophic rhizobacteria are highly adapted to one another. Many facultative or even obligatory endophytes of roots are diazotrophic rhizobacteria. Similar processes to mutualism might also include the rhizospheric indirect effects on plants through nitrogen fixation. A significant portion of the nitrogen fixed by free-living diazotrophic rhizobacteria may be stored as ammonia inside of their cells, blocking direct transfer to the plant. Due to the protozoan predation of the bacteria, which boosted the wheat intake of nitrogen by 20%, it was reported that inoculating wheat with diazotrophic bacteria and protozoa increased the mineralization of fixed nitrogen and its uptake by 65%. Bacteriophages nematodes can mineralize nitrogen up to six times more than protozoa feeding on bacteria. Although the direct delivery of fixed nitrogen to plants seems to be cyclical, the widespread presence of diazotrophic bacteria in the rhizosphere raises the possibility that trophic interactions within the soil food web are more important for plant nitrogen uptake than has previously been acknowledged.

4.13 RHIZOBACTERIA

Rhizobacteria are categorized as either detrimental rhizobacteria (DRB) or plant growth-promoting rhizobacteria (PGPR) depending on whether they positively or negatively affect plant growth. Through a variety of mechanisms, such as the production of phytohormones, nitrogen fixation in the rhizosphere, P solubilization, mycorrhizal function promotion, and regulation of ethylene production in roots, PGPR can directly increase plant growth. Through a variety of mechanisms, such as carbon substrate competition, the sequestration of $Fe3+$ in siderophores that the pathogen cannot access, an increase in the availability of Mn in the soil, the production of antibiotic substances, and the induction of systemic induced resistance in the plant, PGPR can indirectly promote plant growth by suppressing plant pathogens. A lot of research has been done on the possible application of PGPR for the biological control of disease, particularly with pseudomonads because of their high rhizosphere competence. Key mechanisms in the biological control by pseudomonads are the induction of systemic resistance in the plant and the synthesis of antibiotic compounds, particularly phenazines and phloroglucinols. Bacteria in rhizosphere communities function across PGPR to DRB. Rhizobacteria have been shown to have detrimental impacts on a wide variety of crops, pastures, and weed species. DRB has also been linked to yield decreases that are connected to continuous monoculture. The production of phytotoxins, such as HCN, phytotoxic metabolites, and phytohormones, particularly IAA, as well as the inhibition of mycorrhizal fungi, diazotrophs, and PGPR, as well as competition with the plant for nutrients like P and Fe, are some of the mechanisms by which DRB may hinder plant growth. Rhizobacteria were classified as DRB or PGPR. Depending on the host genotype, mycorrhizal status, and environmental factors, the impact of different rhizobacteria isolates on plants can range from growth inhibition to growth promotion. Rhizobacterial communities play a significant role in determining whether the soil is favorable to or suppressive of fungal infections. It is clear that rhizobacteria can negatively impact plant growth while also inhibiting fungal plant diseases due to the occurrence of delicate species-specific bacteria-host interactions in the rhizosphere.

Figure 4.10 *Cochliobolis sativus*

Source: https://upload.wikimedia.org/wikipedia/commons/5/57/Cochliobolus_sativus_-_Lindsey.jpg,*Creative Commons License.*

4.14 COLONIZATION AND BORDER CELLS

Specific regulation of bacterial colonization of the rhizosphere can be achieved by variations in host genotype. For instance, diazotrophic *Bacillus* spp. were found to preferentially colonize a genetically engineered line of wheat(Kennedy et al., 2004). The overall quantity of bacteria on the cultivar roots was either lower or the same as that of other wheat lines, suggesting that the host-specific effect was caused by the quality of the root exudates. Genes on a single chromosome controlled the presence of a rhizoplane microflora in spring wheat that was antagonistic to *Cochliobolis sativus* (Figure 4.10). Many aquatic and parasitic plants whose roots lack a root cap structure do not produce border cells. Border cells in solanaceous plants often have minimal energy, and members of the Brassicaceae family including the model plant Arabidopsis thaliana do not produce viable border cells. The cross-phyla phenomena of border cell development in angiosperms, gymnosperms, and pteridophytes indicate that they have been a component of the rhizosphere since the beginning of root evolution. It was previously believed that border cells served as a protective barrier, lowering friction between the soil and the root surface. Research has shown that border cells can produce enough of these cells to fulfill this function. Nevertheless, border cells experience alterations in size, shape, and gene expression after differentiating. Many border cell-specific proteins are produced as a result of the changed gene expression, and these proteins are primarily exported to the environment. These released proteins can affect rhizosphere microorganisms. The root cap is an important location for root exudation, which influences the early establishment of microorganisms. The release of galactose from their mucus by the extracellular galactosidase that the border cells secreted either stimulated or inhibited the colonization of diverse bacterial species. Border cells can generate certain extracellular signal molecules that have a direct impact on gene expression and microbial development, as well as the attraction of nematodes or pathogenic fungi.

Chapter 5

PHYSIOLOGY OF FUNGI

Contents

5.1 INTRODUCTION

The osmotic gradient is important in a walled cell, like a fungus, to sustain the required internal hydrostatic pressure. The plasma and vacuolar membranes are two key locations. We can refer to the solute movement against their electrochemical potential gradient, active transport when solutes are moved across these two membranes or any other membrane. If the solute is polar, diffusion of the solute across the membrane is also possible. This is made possible by carrier proteins, which help the solute get past the activation energy needed to pass through the lipid component of the membrane. Diffusion can only occur along the gradient of electrochemical potential. However, it is important to keep in mind that active transport would occur if the carrier affinity for the solute on one side of the membrane were significantly higher than on the other, or if the solute transferred via the carrier moved more quickly in one direction than the other. Both of these changes would depend on the cell. There are two categories of active transport: primary and secondary. Reactions involving the donation or reception of electrons or the exchange of primary bonds between distinct chemical groups result in primary active transport or translocation. A chemical group or solute is translocated across the membrane as a result of these processes. Thus, a vectorial process is generated through the utilization of chemical energy. Electron donation or acceptance, or primary bond exchange between distinct chemical groups, is not involved in secondary transport or translocation. Therefore, proteins play a different role in secondary active transport than they do in primary active transport.

5.2 TRANSPORT

Secondary active transport-related proteins do not operate in the same way as enzymes(Geck & Heinz, 1981). They are referred to as porters. The ability of porters to reversibly catalyze vectorial processes is a result of their type of action. Chemiosmotic reactions are those that result in primary active transport. These processes are what propel purely osmotic or secondary active transport reactions. These reactions fall into one of two categories: sym- or antiport, which includes coupled flows, or uniport, which results in exchange diffusion. Only when a solute is charged and there is a sufficient electrical potential differential across the membrane to provide the driving power for such accumulation can a uniport cause solute accumulation. The movement of one solute against its electrochemical potential gradient can result from the flow of another solute, such as protons or sodium, down its electrochemical potential gradient. Because of the free energy in the gradient of the other solute across the membrane, there is active transport of one solute across the membrane. The transfer of charge and any resulting variations in concentrations on either side of the membrane are important factors to take into account when analyzing secondary translocation. Secondary translocation can result in the movement of charge. The extrusion of protons using the free energy of ATP hydrolysis is the principal active transport at the plasma membrane in fungi. There is limited evidence supporting principal active transport pathways. Fungi primarily use a proton to interact with the external medium. Protons are the main solutes that cotransport with other solutes that need to enter or exit fungi, while there are indications that sodium may also accompany solutes like phosphate in translocation processes.

Figure 5.1 *Neurospora crassa*

Source: https://upload.wikimedia.org/wikipedia/commons/d/d5/Oncom_tekstur.jpg,*Creative Commons License.*

5.3 NEUROSPORA CRASSA

Because the H^+-ATPase of *Neurospora crassa* (Figure 5.1) was studied more than that of other fungi, more is known about it than about other fungal enzymes(Palmgren & Morsomme, 2019). Electrophysiological studies have made it possible to describe the proton extrusion process in the case of primary active transport in *N. crassa*, accounting for both the membrane electrical field in which the enzyme resides and the properties of the ATPase in vitro and in membrane vesicles. The reason why *N. crassa* electrophysiological experiments are successful is that the mycelium growing on cellophane on agar has hyphae with a diameter of 10-20 μm. The diameter is wide enough to accommodate the insertion of electrodes for pH measurement and voltage measurement. Electrochemical protons are extruded from the hyphae by the active H^+-ATPase. As a result, the enzyme can be described as functioning as a physiological proton pump. The membrane potential difference is -25 mV if the pump is non-functional. When the pump is working, the potential difference under typical physiological conditions is -270 mV, meaning that the activity of the pump generates about -200 mV of potential. The potential difference can be as low as 60 mV when a mycelium is in 100 mM potassium chloride in a calcium-free medium; but, in the absence of chloride, which tends to depolarize the membrane, the potential difference can be as large as -300 mV. The fact that the loss of the potential and the decomposition of ATP in the mycelium with the addition of sodium azide are equivalent provides evidence that ATP is the substrate for the pump. The proton pump has now been thoroughly described on a molecular and biological level. Contamination from ATPases found in other membranes, such as mitochondrial and vacuolar membranes, is the main issue with isolating the plasma membrane H+-ATPase. A hydropathy plot has made it possible to determine the polypeptide transmembrane orientation. The relationship between the polypeptide structure and function has been further investigated through the use of site-specific agents such as N-ethylmaleimide (sulphydryl groups), phenylglyoxyl (arginine), and N-(ethoxycarbonyl)-2-ethoxy-1,2-dihydroquinoline for carboxyl groups. These investigations are giving us a better understanding of the residues necessary for catalytic activity. Specifically, we can now see the polypeptide transmembrane topography as well as the sections connected to the cytoplasm The enzyme from *N. crassa*, *Saccharomyces cerevisiae*, and *Schizosaccharomyces pombe* is studied. The ATPase of *S. cerevisiae* and *N. crassa* exhibits a 74% sequence identity, with only 238 variations in the amino acid sequences, totaling 918 and 920 residues, respectively, illustrating the extent of the similarity. 105 of these variations are replacements of similar amino acids. With 74 variations in the first 100 residues and 42 in the final 100, the N- and C-terminal regions exhibit the greatest variation.

Nearly all of the knowledge regarding the function of the proton pump about the physiology of a specific fungal species and the regulation of pump activity comes from studies on *N. crassa* and *S. cerevisiae*. The pump causes the buildup or extrusion of other solutes inside the fungus by pumping protons out of the cell, which creates the majority of the electrochemical potential gradient for protons across the plasma membrane. The difference in proton concentration across the membrane, or pH gradient, and the electrical potential gradient, which is mostly produced by the proton pump but not entirely, makes up the electrochemical potential gradient. The accumulation of another solute, supported by a carrier that carries protons down the gradient, or the extrusion of a solute, in this case, antiported with the protons via the same carrier, are both driven by the free energy in the proton electrochemical potential gradient. Researchers demonstrated that the proton gradient's direction has no bearing on the cytoplasmic acidification brought on by respiratory inhibition. It seems that vanadate suppression of pump action has minimal impact on the pH of the cytoplasm.

Even when the pump is stopped, removal of the cyanide causes a pH drop that is restored. It was determined that metabolism plays an equal role in controlling the pH of the cytoplasm. This implies that the proton pump is not the only component engaged in pH regulation. They could be a biological pH-stat, pump activity at vacuolar membranes, and plasma membrane redox systems. Nevertheless, additional data from *S. cerevisiae* suggests that, although a variety of fungus-related mechanisms are responsible for the regulation of cytoplasmic pH, H+-ATPase activity is a crucial component of this complex.

The degree to which the cytoplasmic concentration of ATP is the only factor influencing the activity of the enzyme as a proton pump is another important aspect of H^+-ATPase operation. Studies on the *N. crassa* strain known as poky f provide some of the evidence(Gowdridge, 1956). This strain lacks cytochromes a and b due to deficiency in its mitoribosomes but has large amounts of the alternative oxidase. The ATP in the wild-type decays exponentially. In 1 mM cyanide, the poky f strain exhibits an exponential decrease; nevertheless, there is a recovery to roughly 70% when the mycelial concentration reaches roughly 40% of the initial value.

In poky f, there is depolarization (30-100 mV) followed by mild hyperpolarization, which suggests ongoing pump regulation. In the wild-type, the membrane potential depolarizes exponentially. The pump current-voltage characteristics in poky f demonstrate that the pump proton stoichiometry changes from one to two protons per ATP molecule in the presence of cyanide. The mechanisms that appear to be responsible for the mycelial depolarized membrane, comparatively high ATP levels, and altered pump stoichiometry in the presence of cyanide appear to be the same. Cyclic AMP (cAMP) may be involved. Evidence from wild-type mycelium suggests that changes in membrane potential do not always correspond with changes in intrahyphal ATP concentrations. This suggests that similar to poky f, cAMP must be involved in other forms of metabolic control in addition to ATP concentration. Research conducted on *S. cerevisiae* provides an additional perspective on the topic of controlling plasma membrane H+-ATPase activity regulation.

5.4 SIGNALING METABOLITES

cAMP mediates growth and glucose metabolism and H^+-ATPase activity(Becher dos Passos et al., 1992). Ras-adenylate cyclase signaling pathway is involved only in the shift from the derepressed state where *S. cerevisiae* cells grow on respiratory carbon sources, to the repressed state, where they grow on fermentative carbon sources. A thermosensitive mutant with adenylate cyclase activity defects has a 50% decrease in plasma membrane H^+-ATPase activity in just one hour. The information showed that cAMP functions as a negative effector on the mitochondrion but a positive effector on the plasma membrane's H^+-ATPase. Since the H^+-ATPase of *S. cerevisiae* (Figure 5.2) lacks any putative phosphorylation sites, clusters of arginine residues followed by serine, recognized by such kinases, it does not appear to be the result of a cAMP-dependent protein kinase.

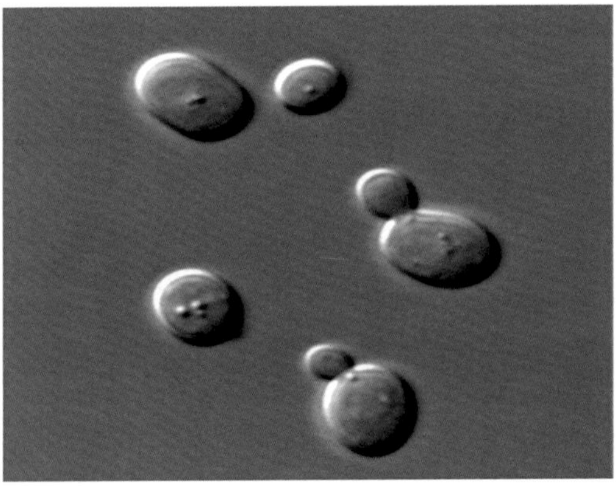

Figure 5.2 *S. cerevisiae*

Source: https://upload.wikimedia.org/wikipedia/commons/d/d9/S_cerevisiae_under_DIC_microscopy.jpg,*Creative Commons License.*

Though there are other possibilities, there is strong evidence that the H^+-ATPase can be phosphorylated in vivo. This phosphorylation may be caused by a protein serine kinase in the plasma membrane. When an enzyme acts on glucose, it phosphorylates specific locations on the enzyme that correspond to where it is located in the plasma membrane and some sites are phosphorylated before the enzyme reaches the membrane. When glucose is removed, there occurs dephosphorylation, and when glucose is added again, there is a quick reversal. The regulatory domain at the C-terminal end of the H^+-ATPase polypeptide is probably another way. No principal active transport systems have been found to exist in the plasma membranes of fungi; however, further research on a large number of additional fungi is required to adequately support this claim. Nonetheless, the discovery that *S. cerevisiae* possesses plasma membrane oxidoreductases raises the possibility that additional metabolic processes at the membrane may be pertinent to the transport mechanisms that take place there(Lemire & Oyedotun, 2002).

The NADH and NADPH cytochrome c reductase (NCR) activity is a crucial component of the plasma membrane oxidoreductase system. Nitrate, oxygen, and semidehydroascorbate have been found to be electron acceptors in flowering plants. Moreover, in flowering plants, the NCR activity merely makes up a small portion of the plasma membrane redox activity. It is believed that the NCR activity plays a significant part in the signal transduction process mediated by indole-3-acetic acid (auxin), wherein the ejection of protons and acidification of the wall results in cell expansion as a primary reaction. It is uncertain where the protons are coming from. They are created by the cytoplasmic metabolic processes or by oxidizing a substrate that contains either NADPH or NADH. The modulation of intracellular pH has been proposed to be associated with the NCR system. Ferricyanide has been demonstrated to lower the cytoplasmic pH of flowering plant cells when it is present, however, there is currently insufficient proof to support a regulatory function in healthy plants. Ferricyanide can be reduced by *S. cerevisiae* cells that are either respiratory-competent or respiratory-deficient. The reduction of ferricyanide results in an enhanced loss of protons from the cells; this process is promoted by ethanol and prevented by the alcohol dehydrogenase inhibitor pyrazole. The oxidation of ethanol regenerates NADH, the internal reductant, with protons being released like that of flowering plant cells. *S. cerevisiae* had a plasma membrane redox system, which was thought to be responsible for K^+-H^+ and K^+-Na^+ exchange. The effects of dyes at concentrations of 100 μM had redox potentials ranging from +290 to -160 mV. The redox potential was roughly + 180 mV in the absence of dye. Increased K^+-H^+ exchange occurred when it was elevated above this value by adding the appropriate dye; the opposite happened when the potential was lowered. Potassium ions in the media inhibit the rate of reduction. The relative concentrations at 50% inhibition were found to be identical to the reciprocals of the transport affinities of the same cations for the alkali cation carrier. Other alkali cations also exhibit an inhibitory effect. Alcohol dehydrogenase lowers the K^+ carrier and transfers protons and electrons to methylene blue when there is no exogenous K^+ present(Zohri et al., 2024). When methylene blue is in its reduced form, alkali cations, especially K^+ carried by the carrier compete with it for the carrier. A fungus growing in its native habitat is subject to a changing environment because of the substratum heterogeneity or the fungal metabolic processes, including proton extrusion. A fungus must compete with other bacteria for resources in the natural environment. The fungal capacity to absorb nutrients is determined by the flow of nutrients into it; the more nutrients that are absorbed, the more competitive the fungus will be if nutrient supplies are few. Therefore, it is thought that the higher growth of mycorrhizal roots in comparison to non-mycorrhizal roots is supported by the increased transport of phosphorus into beech mycorrhizal roots as opposed to that into non-mycorrhizal roots. Because the surface area has not been measured directly but rather has been estimated from fresh and dry weights as well as mean hyphal diameters, the values that have been obtained are, for the most part, just assessments. Surface area-based calculations of flux values are required. Measurements must be done using young mycelium to determine the amount of nutrients flowing into the mycelium.

5.5 NUTRITION

A non-uniform concentration will complicate the calculation of the nutrient flux, and as the particle gets bigger, decreasing oxygen tension within the pellet will directly impact the rate of transport as well as other metabolic properties. Without knowledge about the nutrient content and the pH of the area where the fungus is growing, information regarding the flux of a specific nutrient is useless for evaluating what might be happening in the natural environment. For those nutrients whose flux is influenced by the proton electrochemical gradient in the plasma membrane, understanding the latter is essential. The amount to which nutrients may be transferred from older regions of the mycelium to the spreading hyphal apices to promote growth is another problem with filamentous fungus. The nutrients that are transferred may come from the breakdown of polymers in aged protoplasm as well as those that are absorbed from the external medium. The external medium composition has been solely used to evaluate the ionic composition of the current in fungus. pH variations near the hyphae can also reveal information about how much protons are involved. The circuit that enters the hypha at one location and exits it at another, a considerable distance away, is the fundamental characteristic of the current in fungus. Current is therefore conveyed by the hypha itself. Rhizoids are affixed to a thallus with a spheroid-like form that makes up *Blastocladiella emersonii*. Eventually, the thallus separates to become a sporangium. Along with the production of rhizoids, the *Allomyces macrogymus* zygote also develops apically developing hyphae during germination. It is important to keep in mind that while *Achlya* sp. and *Neurospora crassa* share a similar vegetative development pattern, they belong to different sectors of the fungal kingdom. There have been claims that Achlya is not a real fungus, but rather an Omycetes.

Protons are the important candidates to be the ions carrying the current in all of the fungi that were studied(Monk & Perlin, 1994). The proton conductors affect the orientation of an organism developed in culture when it is exposed to glass fibers covered in insoluble inhibitors or ion exchange resin to which inhibitors are absorbed. Ionophores for calcium and potassium have minimal impact. The membrane potential acts like a potassium diffusion potential when microelectrodes are introduced into the thallus. Potassium ions leave the cell by a route that also transports barium and rubidium ions, but not lithium, sodium, cesium, magnesium, or calcium ions, although the last one is necessary to preserve the potential difference. Anions do not alter the potential in any way. When cells are starved of potassium, the thalli absorb the potassium and release protons in a coordinated manner. The pH gradient that exists across the membrane is eliminated by respiratory ATP synthesis inhibitors. Proton-coupling uncouplers and respiratory ATP generation inhibitors are two methods of preventing potassium inflow. These findings suggest that potassium is absorbed by an active mechanism that involves the exchange of protons for potassium ions. Therefore, the most likely scenario is that protons enter through the thallus and exit through the rhizoids. Potassium ions, not protons, carry the current outwards.

Figure 5.3 *Aspergillus nidulans*

Source: https://upload.wikimedia.org/wikipedia/commons/0/09/Aspergillus_nidulans_wildtype.jpg,*Creative Commons License.*

It is obvious that although glucose and phosphate are necessary for the current generation, phosphate cannot carry the current by itself because it would need to move the ion out of the hyphal apex. Because glucose is repressed at the high concentrations used in the experiments, it is unlikely that a glucose/

proton symport would be active. It will be necessary to have potassium ions for charge balance. The protons are thought to be released into the medium at the hyphal apex by the mitochondria located beneath it. It is thought that the protons carrying the current exit the hyphae in the distal zone through the H^+ -ATPase. Protoplasts are released from the apex of the hyphae first, and then gradually from older regions as a result of the hyphal walls being broken down by enzymes. Apical protoplasts in *Aspergillus nidulans* exhibit a lower K^+:Na^+ ratio than those from earlier locations(Isaac et al., 1986). This is consistent with X-ray microanalytical studies on D. salina hyphae. Therefore, there is proof from research other than electrical current tests indicating the transport properties of the hyphal apex in fungi related to *N. crassa* are different from those of the earlier regions. Patch-clamp investigations provide even more direct evidence for this in the oomycete *Saprolegnia ferax*.

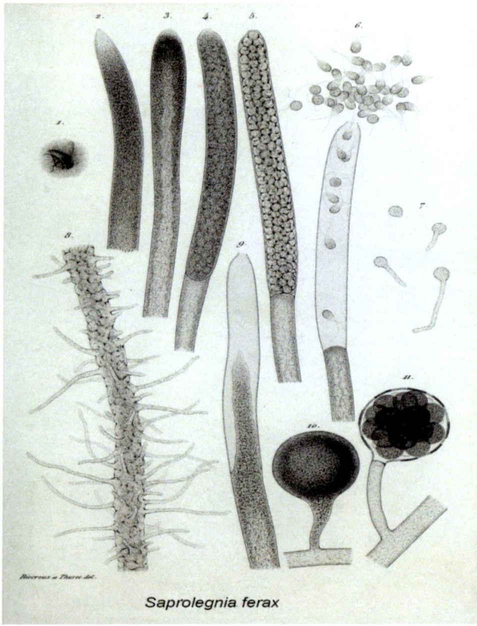

Figure 5.4 *Saprolegnia ferax*

Source: https://upload.wikimedia.org/wikipedia/commons/a/a4/Saprolegnia_ferax_illustr.jpg,*Creative Commons License.*

5.6 HYPHAE DYNAMICS

Concerns that these currents might be essential to the polar expansion of fungal hyphae accounted for most of the research on circulating ion currents in fungi. Researchers agree that there is not enough information to make a firm conclusion on how ion currents and cell polarity are related. Hyphae with inward apical currents in some fungi may occasionally have outward currents. There is no strong correlation between the current density and the length of the peripheral growth. Hyphae can reverse develop, meaning that after growing under low oxygen conditions, they may extend backward, but the current direction stays the same. However, the direction of growth may be related to the flow of protons rather than the net flow of ions. If the current changes without the hyphal extension rate changing, this could be the result of a new flow of other ions while the proton flow stays the same. Regardless of the function of transhyphal currents in hyphal development, research on these currents has undoubtedly produced significant insights into the spatial characteristics of fungal hyphal nutrition uptake. Thus, we now know that amino acids may be taken at the Achlya hyphal apex, and it is possible that *N. crassa* can similarly absorb phosphorus. A portion of the molecular basis for Achlya and other Mastigomycotina members' capacity to exhibit positive chemotropism toward amino acids, particularly methionine, is provided by the existence of proton symport systems in the hyphal apex membrane. If an array of transport proteins for a given amino acid is present in the membrane over the apex and is uniformly distributed, then, this array should be able to identify the location of amino acid sources because the transport proteins closest to the source will be exposed to higher concentrations.

5.7 RHIZOIDS

Rhizoids are recognized to be crucial for *Blastocladiella emersonii* nutrient uptake. Rhizoids are capable of absorbing a wide range of nutrients, and they preferentially absorb phosphate ions and amino acids over thallus. Researchers found that the rhizoids exhibit extensive orientation to gradients of phosphate and a specific blend of amino acids, but not to glutamic acid, glucose, potassium, or calcium. *Allomyces macrogynus* rhizoids (not hyphae) are chemotropic, indicating that these structures are important for nutrient uptake. Fungi are successful heterotrophic organisms because they can break down insoluble material into soluble components to supplement their diet in addition to scavenging nutrients and invading new substrata. Nutrient substrates are solubilized through the secretion of enzymes. However, it has long been unclear where exactly enzyme secretion occurs. The size of the released enzymes is greater than the pore size of the hyphal walls, as measured by molecular exclusion using living hyphae or isolated cell wall fractions. Although isolated walls do not allow polymers with molecular masses higher than 4.75 kDa to penetrate, ribonuclease, the smallest exoenzyme released by *Neurospora crassa*, has a molecular mass of 13.7 kDa(Vega et al., 1991). A porous cell wall at the hyphal apex would facilitate the easier passage of exoenzymes. The structure becomes less porous when the apical wall changes into the lateral wall, which leads to some exoenzymes becoming stuck in transit and adhering to the wall. There is a decrease in the molecular sieving of produced exoenzymes during the germ-tube expansion from spores and early log phase when hyphal apices make up a larger percentage of the total mycelial surface area than in older cultures. Studies conducted on *Aspergillus niger* protein production have provided concrete evidence in support of the theory. The mycelium was cultivated in a thin layer of agarose encased between perforated polycarbonate membranes, allowing for the free flow of produced protein as well as gases, nutrients, and water. The mycelium was fed 35S, and the dispersion of radioactivity in the media allowed for the location of protein secretion.

A hyphal growth from a spore that has germinated follows an exponential curve until the protoplast has doubled in size. Following protoplasmic doubling, a branch/fragment is created, and further branches continue to grow, this process is known as the duplication cycle. Following the initial branch development, the leading hyphae extend linearly. Every hypha is reliant on a supply of precursor materials. The evidence suggests that these substances are enzymes and wall precursors that are carried to the tip of vesicles by contractile elements in the cytoplasm. The peripheral growth zone is defined as the length of the hypha that produces the supply. The branches cover the gap created by the divergence of individual leading hyphae as the colony circumference grows. Regardless of the mechanism, the colony's capacity to evenly distribute its hyphae throughout a homogeneous substratum allows for the best possible utilization of the nutrients present in that substratum. Since the position of the branch is defined by the concentration of nutrients (oxygen dissolved in the substratum), it is the position where the concentration is at its highest about the two leading hyphae. Similarly, the same location will have the least impact on pH due to metabolic activity from the two leading hyphae, as well as the lowest concentration of chemicals emitted by the same hyphae that could influence branch growth. As a result, it appears quite likely that the growing tip of the developing branch needs to be able to sense the solute concentration in the agar such that the development direction is in line with the solute maximum or minimum concentration.

Many research investigations have been conducted on chemotropism in fungus, with the majority focusing on the orientation of spore germ tubes. These investigations have often led to the conclusion that only the Mastigomycotina exhibits a tropic response to nutrients like amino acids, yet research suggests that ectomycorrhizal fungus may be an exception. It has been proposed that in higher fungi, leading hyphae react negatively chemotropically to chemicals produced by individual hyphae because they grow apart and roughly parallel to one another in nutrient-poor conditions. A lot of the settings that researchers have previously employed may have had an impact on the direction of growth due to the oxygen gradient in the medium. Many fungi may be favorably chemotropic to oxygen(Chet & Mitchell, 1976). Bending in the direction of a higher oxygen concentration is caused, in part, by unequal respiratory activity on opposite sides of a hypha. It was proposed that the amino acid proton symport served as the sensing mechanism. It is possible that protons, rather than oxygen, are the key factor influencing chemotropic growth in fungi other than the Mastigomycotina. Growth away from a higher concentration of hydrogen ions (low pH) in the medium resulting from the same metabolic activity may therefore occur concurrently with growth along a gradient of oxygen concentration produced by the same metabolic activity. Protons have an advantage over oxygen as an element in that they can only get through the plasma membrane through a transport protein or porter. As a result, certain sensing agents will be present on the plasma membrane. The cytoplasm must be free of turbulence to prevent gas from diffusing to the mitochondrion to detect a difference in oxygen content between one side of a hypha and the other. Microscopic observations show that this is not the case, with significant cytoplasmic mobility occurring during growth. It is necessary to note that while chemotropism is primarily determined by the concentrations of oxygen and hydrogen ions in the medium, there is currently insufficient data to rule out the possibility that inorganic ions like potassium and phosphate also play a significant role in chemotropism.

There are some cases where solute transport is specifically impacted by compositional changes in the plasma membrane. Researchers demonstrated that variations in the ergosterol content affect the uptake of some amino acids and the efflux of potassium in *S. cerevisiae* cells. Conversely, as *Candida albicans* cells grow in alkanes with varying chain lengths, the uptake of some amino acids is decreased but not of others. It is difficult to interpret these results. Therefore, it is unclear if the impacted transport processes occur at the tonoplast, the plasma membrane, or both membranes. The majority of interpretations also take into account modifications to the environment in which the transport protein operates. Without extensive research, it is impossible to determine whether such interpretations are accurate. Although lipid regulation of transport pathways is considered significant, lipid modulation of plasma membrane ATPases in plants, mammals, and fungi revealed that the mechanisms involved in this process are debatable. Lipid fluidity may have an impact or protein-lipid binding may have an impact. To determine the degree to which the lipid composition of the plasma membrane influences its functioning, other methods will be required. Studies on the operation of transport proteins in synthetic lipid membranes with known compositions would be important. A nystatin-resistant strain of Rhodotorula gracilis has less ergosterol than the wild-type. Although there is no change in the proton electrochemical potential gradient, the mutant cell ability to transport xylose is diminished. It has been demonstrated that nystatin acts only at the plasma membrane in R. gracilis cells. Therefore, it is most likely that the mutation is affecting this membrane. Compared to wild-type cells, the mutant cells seem to have less passive permeability to potassium and proton ions. As a result, in contrast to wild-type cells, mutant cell membrane potential is mostly unaffected by the extracellular potassium concentration and the medium pH. Since the carbonyl p-trifluoromethoxyphenyl hydrazone (CCCP), an ionophore for protons, increases the flux of both protons and potassium, the efflux is always high enough to compensate for the influx of protons caused by CCCP, the governing process appears to be the passive flux of protons. Therefore, it would seem that the reduced velocity of charge compensation required for the proton/sugar symport to work well is the cause of the decreased sugar uptake.

5.8 LIPID MUTANTS

Zygosaccharomyces rouxii has been shown to include nystatin-resistant mutants with varied sterol contents and altered sterol composition(Ushio et al., 1991). The traits of the mutants were distinct from those of the wild type. Specifically, in the presence of 1.47 M sodium chloride, two mutants with changed sterol composition grew less and leaked more glycerol into the media than did wild-type cells. Leakage has not been described as a process. These investigations on the nystatin-resistant mutants of *Z. rouxii* and *S. cerevisiae* provide credibility to the theory that identifying and isolating mutants with lipid membrane defects could be a useful method for understanding the function of the plasma membrane. The mitochondrial membrane seems to be the most susceptible to alterations in the lipid content. Thus, variations in the transport characteristics of lipid mutants may result from the cells' ability to produce energy. *S. cerevisiae* cells grown anaerobically will accept a wide range of sterols and fatty acids as replacements for the naturally occurring 16–18-carbon fatty acids and ergosterol. It is quite easy to create glucose-grown cells with a broad variety of phospholipid compositions by simply changing the growth medium properties. The fact that the lipid content of the plasma membrane and tonoplast can be altered to such a great extent in cells where mitochondrial function is not necessary suggests that these membranes can be relatively flexible without compromising their ability to perform their intended functions. As a result, one must anticipate that lipid mutants will probably exhibit modest changes to the properties of transport processes at the plasma membrane or tonoplast. This is what has been noted regarding the impact on amino acid transport by changing the lipid composition of yeast cells.

5.9 VACUOLE AND TONOPLAST STUDIES

Large algal coenocytes like Chara and Nitella have vacuoles that are large enough to allow for the easy removal of the vacuolar sap for examination. The vacuole contributes to the cell osmotic potential in green plant cells due to its large size in relation to the cytoplasm. It has been possible to determine the fluxes of the main ions across the vacuolar membrane (tonoplast) by analyzing the flux of radiotracers into and out of the coenocytes. The driving force on any given major ion can be ascertained by inserting an electrode into the vacuole, which makes it easy to estimate the cytoplasmic and vacuolar concentrations of these ions as well as the potential difference across the tonoplast. One vacuole does not occur frequently in fungus. There does not seem to be a single vacuole in actively growing hyphae; in fact, multiple types of vacuoles may exist based on how various vacuoles react to stains. The cytological characteristics of vacuolar organization in hyphae have not received the necessary level of attention in research. It is improbable to ascertain whether the vacuoles visible under a microscope are linked or separate entities. Vacuoles in *Saccharomyces cerevisiae* can either combine to create a single large unit or be scattered as multiple smaller units in the cytoplasm, depending on the cellular and cultural conditions(Rieder & Emr, 2000). Throughout the cell cycle, the vacuolar volume varies in tandem with the cell volume. A comparison between the fungus is possible based on the potassium and sodium values. The extremely low quantities of the two cations in the *N. crassa* isolated vacuoles raise concerns at this point because they may indicate cation loss during the isolation process. However, the extremely high potassium content found in *S. carlsbergensis* vacuoles points to the ion entry into the vacuole during differential extraction.

The tonoplast is far more elastic than the plasma membrane; this increased elasticity most likely enables the tonoplast to withstand protoplast rupture under hypoosmotic conditions. It is highly likely that the differing chemical compositions of the two membranes account for the variation in elasticity between them. The tonoplast's resistance to nystatin may be explained by the low sterol concentration. Compared to the plasma membrane, the tonoplast contains less carbohydrates. Consequently, concanavalin A does not adhere to the outer surface of isolated vacuoles, but it does bind to the plasma membrane's outer surface. The tonoplast is approximately 10 times more resistant to polycationic molecules like cytochrome c, polylysine, and DEAE-dextran than the plasma membrane is. The measurement of the solute concentrations in the cytoplasm and vacuoles has been made possible by the plasma membrane's increased sensitivity to these substances. It should be noted that while protoplasts are stable in these circumstances, vacuoles are extremely susceptible to pH values below 6.5, high potassium chloride concentrations, and relatively modest quantities (5 mM) of calcium or magnesium. In slightly alkaline (pH 7.5–8.0) environments, vacuoles exhibit greater stability compared to protoplasts. In addition to having a H+-ATPase, green plant cells also have pyrophosphatase in their tonoplasts. This enzyme causes a pH gradient and potential difference to be created across the membrane. It has been confirmed that *S. cerevisiae* contains a pyrophosphatase in its protoplasm; this enzyme accounts for 90% of the total pyrophosphatase activity. It has been demonstrated that a small portion of that pyrophosphatase is connected to vacuoles. Pyrophosphate had no discernible influence on the membrane potential.

Chapter 6

MICROBIAL PHYSIOLOGY AND PLANT STRESS

Contents

6.1 INTRODUCTION

Plant-associated soil microbes invade the rhizosphere and impact plant development and metabolism to a large extent, contributing significantly to plant function. Microorganisms are helpful for the mineralization of the soil and the breakdown of organic materials. They also keep the biogeochemical cycle and soil ecology intact. In the rhizosphere, the soil microflora and microfauna experience diverse biotic and abiotic stresses and have synergistic or antagonistic effects on plants. A gene pool involved in microorganism-plant interactions is formed by soil bacteria. Bacteria, algae, protozoa, fungi, viruses, and multicellular animal parasites are the main groups of soil microorganisms. The interconnections between soil physicochemical qualities and environmental factors impact the activities of soil microorganisms. All forms of soil contain bacteria, which are essential to fix nitrogen in the atmosphere. Even though the majority of microflora and microfauna are pathogenic, they have a lot of beneficial interactions with the soil. 'Plant microbiome' refers to the collective colonies of microorganisms associated with plants. The study of small soil organisms, their roles, and how they affect soil qualities is known as soil microbiology.

6.2 SOIL MICROBES

In the rhizosphere, soil microorganisms are engaged in a variety of stress situations. The soil serves as a significant organic carbon sink, and soil microbes convert organic matter into plant nutrients. Through their exudates, which contain sugars, amino acids, flavonoids, fatty acids, and aliphatic acids that attract helpful bacteria while eliminating harmful microbes, plants fix carbon into the soil. They are essential to the soil fertility and ecological support. Microorganisms have different sizes and shapes. These have an impact on the biological processes in the soil that are vital to plant and animal life. Soil microbes break down organic residues that are supplied to the soil from different sources, resulting in the formation of humus, which is a complex organic compound with a black color. Water and carbon dioxide are produced when different kinds of microbes break down humus residues. During mineralization, nutritional components contained in organic residues are released in forms that are useful for developing plants. The soil environment has complex biological communities on the planet and accounts for an even greater proportion of biodiversity than tropical forests. Some techniques allowed for the isolation of a wide variety of soil microorganisms, including green algae, Lactobacillus species, Acetobacter, aerobic nitrogen-fixing bacteria (*Azotobacter chrococcum*), and yeast Schizosaccharomyces as well as others.

Beijerinck identified *Bacillus radicicola* or Rhizobium, which causes legumes to nodulate to fix nitrogen(Velázquez et al., 2019). The sulfate-reducing bacteria, *Thiobacillus denitrificans* utilize sulfate as an electron acceptor rather than oxygen. The sulfur-oxidizing bacteria Beggiatoa are chemoautotrophs that utilize CO_2 as a carbon source and inorganic H2S as an energy source.

6.3 RHIZOSPHERE DYNAMICS

In the rhizosphere, a wide variety of soil microorganisms are engaged in different stress situations. Because of their varied genetic composition, soil microbes interact with biotic and abiotic stimuli. They encounter stressors related to temperature, water potential, and nutrients. Many stress response mechanisms of microorganisms interact with one another and play a significant role in the pathogenicity of the organisms. A vast diversity of soil microorganisms responds to different stresses. Escherichia coli bacterial cell wall aids in the conservation, adaptation, and defense of bacteria against stressful environments. Through microbial activity-mediated breakdown, mineralization, and nitrogen fixation activities, soil stress conditions impact soil fertility. Soil microbial activity is significantly impacted by temperature and moisture stress. The fungal cell wall component creates a cell surface epitope in response to changes in the soil environment, shielding the cell from the host immune response. To adapt their metabolic processes to drought stress, soil microorganisms change or adapt. Soil microorganisms change the character of proteins in response to stress conditions by causing genetic alterations (mutations). Antibiotic metabolites are produced by fungi and bacteria to combat pathogenic microorganisms. Microorganisms use physiological acclimation mechanisms to undergo diverse modifications in response to environmental stresses in soil, enabling them to survive. Through a variety of processes, soil microbes affect plant growth and metabolic reactions. The physical environment of the soil is altered by soil abiotic factors like freezing and thawing, drying and rewetting, water percolation, and oxic and anoxic conditions. These factors also have an impact on microbial population and activity. For plants to endure harsh weather, the rhizosphere microbiome is crucial. In saline soil, it was demonstrated that soil microorganisms might sustain plant growth under conditions of osmotic drought stress. Under saline stress conditions, the dominant bacterial genus Bacillus has been isolated.

At various phases of plant root growth, the activity of soil microorganisms varies with respect to time and space. In the rhizosphere, microbial genes are involved in secretions, stress responses, and nutrition uptake. Increased temperature and moisture variations have a significant impact on the population and functions of microorganisms. Grazing alters the diversity and composition of plant species, increases soil carbon loss, and has an impact on the soilborne microbial population(Dignam et al., 2016). Enzyme activity in the rhizosphere and soil microbial biomass are impacted by drought stress. The impact of tillage on soil microbial activity and carbon dynamics during the decomposition process was observed. Due to limited organic matter inputs caused by moisture stress in the soil, carbon reserves in dry areas often only reach half of those in soils in moist climatic conditions. Rainfall episodes set off moisture changes that lead to stress, and stress has a significant impact on the respiration of soil microbes and the mineralization of nutrients. Microfauna often tolerates stress sooner than bacteria and fungi, and the decrease in moisture content in soil decomposers activity lowered their activity. Compared to fungus, bacteria are more sensitive. Polycyclic aromatic hydrocarbons have an indirect effect on the microbial population in legumes, which lowers nitrogen fixation in the root(Kawasaki et al., 2012). The microbial population and plant growth are impacted by some abiotic variables, including pH and hazardous metals. Plants can withstand high levels of stress due to the rhizosphere microbiomes. Through processes like biofertilization, stimulating plant root growth, rhizoremediation, controlling abiotic stresses, and disease control, soil microorganisms in the rhizosphere of plants encourage plant growth against a variety of pathogens. Antibiosis is the primary defense mechanism used by rhizosphere bacteria against plant diseases. Fungi are also producers of antibiotic compounds in the rhizosphere, in addition to bacteria. The majority of biocontrol strains, which include fungi and bacteria, generate several antibiotic compounds with varying levels of antimicrobial activity. It has been demonstrated that the soil microbes promote plant growth in a range of soil types. The activity of soil microorganisms varies depending on the stage of plant root growth and the particular time and place. A gene pool of several microbial species in the rhizosphere shields microorganisms against severe stressors. Many short RNAs (sRNAs) that build up in plant tissues have a role in controlling how plants react to different bacteria during infections.

6.4 PLANT-MICROBE INTERACTIONS

At varying depths, the relationships between microbes and plant roots are also diverse. The majority of algae are located in the uppermost layer, while certain blue-green algae are associated with roots that generate organic compounds. Plant roots are shielded or protected from a variety of diseases by a vast community of soil microbes both before and during primary infection. Antimicrobial substances are produced by certain species of Trichoderma. Multiple antibacterial chemicals are produced by some bacterial and fungal species. These changes in biodiversity affect ecosystem processes and modify how resilient ecosystems are to changes in the environment. Microorganisms play a crucial part in the global fluxes of biogenic gases, which include carbon dioxide, methane, and nitrous oxide. Because these include intricate interactions between microorganisms and other biotic and abiotic factors, the processes by which microorganisms regulate the terrestrial greenhouse gas flux are not well understood. In cold temperature conditions, soil microorganisms and nematode biodiversity are impacted by climate change. Plant wastes are the main source of organic carbon. Soil fauna and microorganisms decompose soil organic matter, resulting in losses to soil carbon stocks. Soil carbon stocks are exported through erosion and leaching. In arid and semiarid locations, organic matter inputs were severely restricted by moisture stress in the soil. Since soil organic matter serves as both a nutrient sink and a medium for nutrient cycling, it is important to assess soil fertility. The moisture content of soil organic matter is directly impacted by changes in climate, which in turn impacts the physical, chemical, and biological properties of the soil. Methanotrophs, microorganisms that oxidize carbon molecules, need copper to oxidize methane(Leak & Dalton, 1986). They also have great potential for bioremediation of soil containing copper, chemical biotransformations, and the production of bioenergy. The main biological method used by nature to reduce atmospheric concentrations of the powerful greenhouse gas methane is methane monooxygenases. Important carbon and nitrogen cycles in ecosystems are mediated by soil microbial populations. Many ecological processes, including the storage of carbon in soil and vegetation, are regulated by the nutrient cycles (nitrogen, phosphorus, and sulfur). The availability of nutrients in soil, microbial biomass, and nematode density are all impacted by climate change. The composition of microbial diversity is altered by soil management methods that involve deforestation, the transition of perennial plant species to annual species, the effects of heavy grazing on soil carbon, and temperature variations in the soil.

6.5 SOIL MICROFLORA

There important soil microflora are bacteria, fungi, actinomycetes, and algae. The bacterial group is categorized as autotrophic (Nitrosomonas, Nitrobacter, sulfur oxidizers) and heterotrophic (symbiotic and nonsymbiotic nitrogen fixers, ammonifiers, cellulose decomposers, denitrifiers). The breakdown of soil organic matter provides carbon and energy for the heterotrophic soil microbes. Protozoa, nematodes, and the majority of bacteria, fungi, and actinomycetes are important. In aerated soils, fungi are the second most available microbial biomass, after bacteria. The fungal population varies between 2×10^4 and 1×10^6 fungal propagules, which represent a portion of the fungus per gram of soil; typically, basidiomycetous fungi account for more than half of the total fungal biomass. Compared to anaerobic bacteria, the population of aerobic bacteria is substantially larger. Pseudomonas and Bacillus bacteria break down complex organic molecules. Important members of the soil microbial community are the coryneforms, nocardioforms, and filamentous bacteria, also known as actinomycetes. These are a large category of filamentous, Gram-positive bacteria that resemble mold. The majority of actinomycetes are found in moist soils with high humus content. As a consequence of their metabolic operations, they release nutrients into the soil solution. Actinomycetes fix nitrogen that is available to plants when they associate with certain leguminous plants. The actinomycetes secrete an exudate (sticky) that contains polysaccharides, which aids in the agglomeration of soil particles. Actinorhizal refers to the relationship between actinomycetes and plant roots. These filamentous bacteria fix nitrogen in nodules. In symbiotic vesicles, nitrogen fixation takes place while the anaerobic conditions required for nitrogenase are maintained. Alnus (Betulaceae), Ceanothus (Rhamnaceae), and other plants form nodules that are the size of baseballs, while Casuarina (Casuarinaceae) form nodules the size of soccer balls. The genus Frankia is a well-known actinomycete that develops nodules in a variety of non-leguminous trees and shrubs that belong to eight families of angiosperms.

Figure 6.1 Armillaria

Source: https://upload.wikimedia.org/wikipedia/commons/8/84/Armillaria_mellea%2C_Honey_Fungus%2C_UK_1.jpg,*Creative Commons License.*

When actinorhizal nodules are opened, they reveal a pink color that is caused by anthocyanins rather than leghemoglobin, which is the case in legumes. Actinomycetes play a significant role in the breakdown of organic matter in the soil and aid in the mineralization of nutrients. They are a source of new bioactive metabolites and antibiotics. Actinomycetes are responsible for producing 70%

of 2200 secondary metabolites known to exist in microorganisms. Researchers reported that the antifungal activity of actinomycetes originated from soil. An actinomycetes strain (*Streptomyces hygroscopicus subsp. ossamyceticus*) with antibiotic activity was identified from the desert soil(Selvameenal et al., 2009). This strain produces a yellow pigment. Using acidified media (pH 5.5), the diversity and characteristics of plant-beneficial acidophilic actinomycetes have been isolated from soil. Wetland soil, saline soil, and marine, brackish, and terrestrial sediments have actinomycetes. It has been determined which Streptomyces species in the soil exhibit antibacterial action.

6.6 FUNGI

One of the largest and oldest living organisms that occupy at least 15 hectares, weigh more than 1000 kg, and have remained genetically stable for more than 1500 years is the facultative parasite of many roots, Armillaria, (Figure 6.1) a basidiomycetous fungus(Hadibarata & Kristanti, 2013). The microscopic, single-celled to multicellular organisms known as fungi typically develop hyphae, which are long, thread-like structures. Certain types of soil fungi live independently, while others form symbiotic relationships (mycorrhizae) with plant roots. The majority of fungus species found in soil are anaerobic Deuteromycetes, Zygomycetes (Mucorales), and Ascomycetes, with a smaller number belonging to Chytridiomycetes and Oomycetes. They are crucial to nutrient cycling and the prevention of disease. More than 690 species belonging to 170 genera have been found in soil. Soils include a wide variety of mold genera, with Penicillium, Mucor, Fusarium, and Aspergillus being the most common. Because they lack chlorophyll pigments, all fungal species rely on the organic matter in the soil for their energy and carbon needs. In the terrestrial environment, fungi break down plant cell wall constituent lignin as well as cellulose, hemicelluloses, pectins, and starch. Based on how they obtain their food, soil fungi can be classified into three categories: saprophytic, symbiotic, and parasitic fungi. The decaying organic matter in the soil is broken down by the saprophytic fungi, which also immobilize proteins and various nutrients like phosphate and nitrogen. Certain fungal species (mycorrhizal fungi) have a symbiotic relationship with plant roots. Mycorrhizae refers to an important fungus community that forms symbiotic relationships with deeper plant roots in the soil. Mycorrhizae obtain their energy sources from plant roots, which are carbohydrates, and expand the surface area around plant roots to the increased availability of nutrients. Root-associated fungus population is influenced by topography and soil moisture. Ectomycorrhiza and Endomycorrhiza are the two primary kinds of mycorrhizal interactions. Many fungal species that form a fungal mantle to cover the outside of plant roots are part of the ectomycorrhizal group. These fungi enter the roots and grow around the cortical cells, but they do not break through the cell walls. The endomycorrhizal group invades and penetrates the cortical cell walls to produce hyphal masses within the cells. The most widely distributed of them are known as vesicular-arbuscular mycorrhizae (VAM).

6.7 NEMATODES

Plant nutrients and other products are stored in the structures known as vesicles. Nematodes are microscopic organisms ranging in size from 2000 μm and can be found in many types of soil(Neher, 2010). There are reported to be over 1000 species of soil nematodes. They eat fungus and bacteria and live freely in the soil. The parasitic fungi inhibit the growth of plants. Nematodes are mostly saprophytes, meaning they feed on dead organic matter; however, some of them can also feed on other nematodes, bacteria, protozoa, algae, etc. Certain Heterodera species can infiltrate plant roots. They seriously damage crops, particularly vegetables. The 20 cm-long Mermithidae nematodes, which are parasites of some arthropods like locusts, are widely distributed in tropical soils. Where a moist layer envelops the soil particles, nematodes can pass through the soil. They inhabit the voids left by wet soil particles. In hot and dry weather, they go into dormancy, but they quickly return to activity as soon as water becomes available.

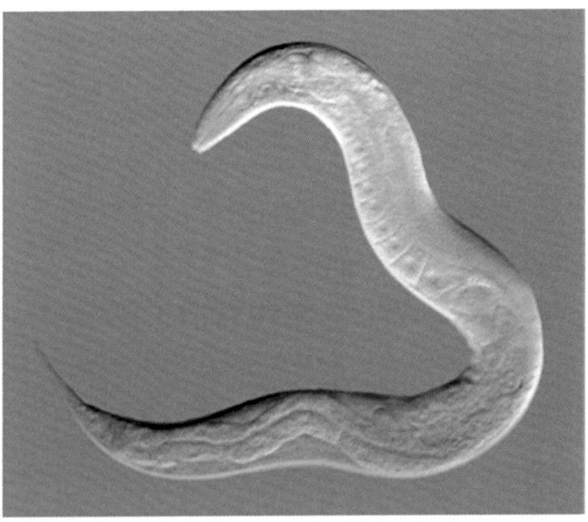

Figure 6.2 Soil nematode

Source: https://upload.wikimedia.org/wikipedia/commons/6/6a/CelegansGoldsteinLabUNC.jpg,*Creative Commons License.*

The allelopathic impact of marigold (Tagetes spp.) produces essential oil that reduces the number of nematodes in the soil by creating a Nematode-suppressive environment. Certain bacterial groups can also inhibit nematode activity through antagonistic effects. Soil nematode (Figure 6.2) populations are decreased by soil solarization and actinomycetes found in organic matter of marigold-rich soil. The most basic type of animal life is protozoa, which are microscopic and single-celled. They are larger than bacteria, mostly feed on other protozoa and bacteria, and have an impact on the pace of mineralization, the activity of microorganisms, and the fixation of nitrogen in soil. Amoebas and flagellated protozoa are primarily constrained to the water-filled pores in the soil because they depend on it for food and mobility. By developing resistant cysts, they are able to tolerate the soil drying out and other unfavorable circumstances.

The application and utilization of microorganisms of agricultural importance will in fact be necessary for global agriculture and enhanced production. To minimize the use of chemicals and enhance crop performance, efficient microorganisms are essential to integrated management techniques including biotic and abiotic stress management and nutrient management. However, there is pressure to use the microbiome to boost crop yields and decrease losses from environmental stresses due to rising population and high food demand. A variety of abiotic stresses affect crop plants including drought, high temperatures, salinity, nutrient deficiencies and flooding which are projected to intensify due to climatic changes. Abiotic stressors pose a serious risk to the productivity of agriculture. One of the main causes of these abiotic stresses is water deficit brought on by weather changes. The most important abiotic stressor is drought, which has a negative impact on crop plant productivity. Even with a short-term water imbalance, plant growth and development are negatively impacted. For instance, drought alone affects up to 45% of the agricultural land, which is characterized by a large human population and growing food demands. Overall impacts of drought on plant growth and development have been documented in a number of studies. Nevertheless, it is challenging to comprehend the damage done to plants at the cellular and molecular levels when there is a water deficit.

6.8 SALINITY

In dry and semiarid regions with low annual precipitation that rely on irrigation for agriculture, salinity restricts agricultural productivity(Jacobsen et al., 2012). Elevated concentrations of salt ions in soil hinder roots' ability to absorb water, which eventually leads to the buildup of harmful salt ions inside plant cells. Plants can deal with issues like low Na^+ concentration by using ATP-dependent ion pumps to actively maintain translocation into vacuolar compartments. However, osmotic stress is brought on by excess NaCl in irrigation water. Any disruption of intracellular ion homeostasis can have detrimental effects. These include disruptions to cell signaling pathways, which can result in the production of proteins, osmotically active metabolites, nutritional deficiencies, assimilation problems, decreased cell division and expansion, genotoxicity, and certain enzymes that scavenge free radicals and regulate ion and water flux. Ultimately, these disruptions impair optimal growth and development. Stressors that are unrelated to one another, like salt, drought, and high temperatures, can cause a similar set of reactions in plants by triggering the same or related pathways. For all of these stresses, a general response has been reported in many crop plants, including compatible solute accumulation and the synthesis of stress proteins and antioxidants at the cellular level. Because our knowledge of plant stress tolerance mechanisms is still lacking, attempts to improve plant performance under abiotic stress have not yielded much progress. To increase stress tolerance, many groups have investigated the intricate processes behind stress response and adaptation, including stress signaling, modifying metabolism, and reprogramming gene expression. However, the production of stress-tolerant plants is a practical strategy for agricultural and environmental sustainability, and it is necessary to meet the rising need for high-quality food. Nevertheless, appropriate tools to control crop productivity in a stressful environment are absent from existing breeding techniques. On the other hand, creating stress-tolerant crop plants can be greatly aided by genetic engineering. It is possible to increase crop plant stress tolerance by combining transgenic techniques with existing breeding techniques. To create stress-tolerant plants, current transgenic approaches aim to transfer one or more different genes involved in multiple pathways to the target plant. These genes include regulatory transcription factors, compatible solutes/osmoprotectants (proline, glycine betaine, polyamines), and proteins (LEA, heat shock, aquaporin. The identification of important genes and their use in transgenic crops with enhanced stress tolerance without yield loss has been and remains the bottleneck of transgenic techniques.

6.9 TOOLS FOR PLANT MICROBIOME ANALYSIS

Using advanced microscopic techniques in conjunction with molecular and omics tools, many important and unexpected discoveries about the plant-associated microbiome have been made over the past decades. Many different agriculturally significant microbiomes have been widely used to boost plant growth and control disease. Beneficial microbiomes associated with plants are predicted to provide a substantial contribution to reducing abiotic stressors through a number of methods(Hussain et al., 2018). Crop plants that are subjected to environmental stressors must find unconventional ways to survive, like using microbiomes. Enhancing microbial characteristics that are advantageous to plants, the ecosystem, or both presents a possible path toward the creation of sustainable agriculture in the future. Under diverse biotic and abiotic stressors, microbial collection and utilization can be an invaluable tool and important determinant in controlling plant health and productivity. Global agriculture would be able to support the next generation if beneficial microbiomes that improve plant resistance to abiotic stress were identified, characterized, and used. The plant-associated microbiome can affect growth and tolerance to biotic and abiotic stresses, among other plant traits

6.10 SOLUTIONS FOR DROUGHT STRESS

Global agriculture is seriously threatened by drought stress. It is not known how the plant-associated microbiome helps plants adapt to drought stress. Researchers found that the ability to promote plant growth is stress-dependent. A pepper plant that was inoculated with specific strains under both irrigated and drought-prone conditions showed signs of stress-dependent plant growth-promoting behavior by boosting photosynthesis in drought-prone grapevine and increasing biomass and length of shoots and leaves. This had a positive impact on drought-sensitive rootstock. Some findings show that the bacteria considerably aided in plant drought adaptation by promoting stress-induced plant growth. Through 1-aminocyclopropane-1-carboxylate (ACC) deaminase, PGPR, such as *Achromobacter piechaudii*, improves drought stress tolerance in pepper and tomato. The breakdown of plant ACC by bacterial ACC, however, would prevent the synthesis of ethylene, which would ultimately reduce plant stress and allow for regular plant growth. The plant performance under drought stress in the early stages of stress development has been evaluated using volatiles from bacterially primed drought-stressed wheat seedlings. When faced with a water deficit, wheat inoculated with *Burkholderia phytofirmans* showed improved water usage efficiency, higher chlorophyll content, and higher grain production than the control. Under drought conditions, three bacterial strains, *Pseudomonas putida*, Pseudomonas sp., and *Bacillus megaterium*, isolated from water-stressed soil, encouraged plant development. Researchers found that Pseudomonas sp. inoculation increased plant biomass and survival of sunflower seedlings under drought stress. The root-adhering soil can probably be effectively colonized by bacteria, leading to soil aggregates and eventually improved stress tolerance. Under conditions of water deficiency, maize plants inoculated with Pseudomonas strain displayed enhanced compatible solutes and antioxidants(Saleem et al., 2021).

Microbes isolated from the roots of plants grown in extremely dry conditions enhanced the growth of another host species grown in conditions comparable to those of the tomato, grapevine, olive, and pepper plants. To induce antioxidant catalase activity in lettuce under extreme drought conditions,

Researchers inoculated the plant with *Pseudomonas mendocina* and arbuscular mycorrhizal fungus (*Glomus intraradices*). This suggests that microbes may be used to reduce oxidative stress. Likewise, the build-up of 14-3-3 protein, in conjunction with glutathione and ascorbate, has been crucial in preserving plant metabolic processes and providing drought resistance. By increasing their water contents, root biomass, and N and P contents, lavender plants inoculated with *Glomus intraradices* and *Glomus* sp. strain exhibited significant drought resistance(Marulanda et al., 2007). Soil reclamation and management techniques are two methods that have been linked to addressing salinity in addition to molecular biotechnology. To control salinity, these approaches are costly and not necessarily viable or sustainable. On the other hand, there is increasing momentum towards the use of naturally occurring bacteria that promote plant growth as inoculants for crop plants that are grown on land affected by salt. Microbial communities have been linked to improved plant health and increased productivity in unfavorable environmental conditions.

Microorganisms living in areas subjected to repeated stress conditions acquired adaptive tolerance features and could be useful as stress-induced plant growth boosters. Microbes tolerant to salt stress in the soil flourish and exhibit characteristics that aid plants in withstanding elevated salinity levels. Genes, hormones, and proteins, including siderophores, gibberellin, indole-3-acetic acid (IAA), and nifH were linked to this tolerance. Similarly, by decreasing ethylene synthesis through ACC deaminase activity, halotolerant bacterial strains improved plant development under salinity stress. Some alternatives include the halotolerant diazotrophic bacteria that exhibit characteristics including phosphate solubilization, ACC deaminase activity, and IAA that were isolated from the roots of Salicornia brachiate, an extreme halophyte. Canola, pepper, tomato, bean, wheat, and lettuce have all shown enhanced growth when exposed to plant-associated microbiomes. Arbuscular mycorrhizal fungi or AMF boosted host plant tolerance to salinity stress. In comparison to uninoculated controls, co-inoculation of AMF plants with Glomus sp. had improved growth in saline soils, due to increased phosphate and decreased Na^+ content in shoots. AMF treatment has decreased NaCl concentration, enhanced osmoregulation by proline accumulation, and improved salt tolerance in maize, mungbean, clover, tomato, and cucumber due to P accumulation. Nonetheless, more study is required to determine whether bacterial and AM species may produce osmoprotectants and protective proteins. Plants that are stressed out can easily attract a variety of bacterial strains, which has extensive effects on plants that are stressed out by salt. Induced systemic tolerance is the collective phrase used to describe this phenomenon. As per the Intergovernmental Panel on Climate Change or IPCC, climate change is expected to raise global temperatures by 1.8–3.6 °C by the end of this century(Maliva, 2021). Both crop productivity and microbial colonization, which cause significant cellular damage like protein aggregation and breakdown, are greatly hindered by high temperatures. Heat shock proteins (HSPs) are a specific class of polypeptides that are produced by all organisms in response to high temperatures.

6.11 STRESS ADAPTATION

Stress adaptation in microorganisms is a complicated regulatory process that may involve several genes. As a result, the microbes adapt in diverse ways to the stress. Certain microorganisms are more effective at high temperatures, and these microbes may be crucial for crop plants. In the drought-stressed rhizosphere of chickpea, researchers identified a *P. putida* strain, which showed thermotolerance and was linked to thick biofilm development and the overexpression of the stress sigma factor. To withstand stress, some bacterial strains produce exopolysaccharides (EPS), which have water-holding properties and are essential for the development of biofilms and stress tolerance.

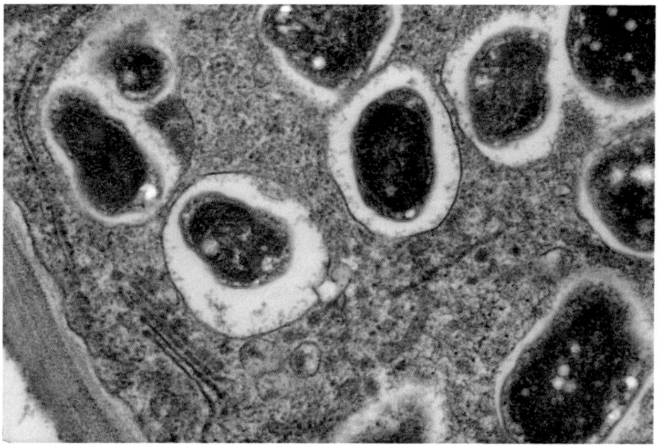

Figure 6.3 *Bradyrhizobium japonicum*

Source: https://upload.wikimedia.org/wikipedia/commons/a/ac/Root-nodule01.jpg,*Creative Commons License.*

Through increased physiological and metabolic efficiency, sorghum seedlings inoculated with the Pseudomonas strain exhibited higher tolerance to heat stress, suggesting a special interaction of inducible proteins in heat tolerant microorganisms. Because low temperatures negatively impact plant growth and development, it is a significant factor limiting crop yield. B. phytofirmans inoculation improved the grapevine resistance to grey mold and low non-freezing temperatures. Endophyte inoculation produced a quick build-up of stress-related proteins and metabolites, resulting in a more effective resistance to low temperatures and suggesting that plants had benefited from priming. Similarly, researchers observed that at low temperatures (4°C), grapevine roots inoculated with B. phytofirmans showed improved root growth, increased plant biomass, and increased physiological activity. In comparison to uninoculated control plantlets, inoculated plantlets produced high proline, starch, and phenolic levels, which improved the ability of grapevine plantlets to withstand low temperatures.

However, inoculating soybeans with both *Bradyrhizobium japonicum*(Figure 6.3) and *Serratia proteamaculans* led to quicker growth at 15 °C. Normally, low temperature suppresses soybean symbiotic activities (nodule infection and nitrogen fixation). Wheat seedlings infected with Pseudomonas sp. exhibited enhanced root and shoot lengths, total phenolics, chlorophyll, and amino acid contents, as well as dry root/shoot biomass. Moreover, inoculated wheat seedlings showed decreased Na+/K+ ratio and electrolyte leakage, as well as increased physiologically accessible iron, anthocyanins, proline, protein, and relative water contents, all of which contributed to their improved cold tolerance. Many bacterial strains have been investigated to improve plant resistance to cold stress.

Since heavy metals cannot be broken down, heavy metal contamination brought on by rising industrialization has drawn attention recently. Many physiochemical and biological methods that were designed to eliminate pollutants have failed because they are costly, hazardous to the environment, and unpopular with the general population. With a high level of public acceptance, phytoremediation, the use of plants to eliminate toxins from soil, is a low-cost, environmentally friendly method. The use of plant-associated microbiomes is another effective option. Microbial activity in the rhizosphere enhances plant metal uptake by changing the mobility and bioavailability of metals. Plant-associated microbiomes produce a variety of substances that aid in plant growth in heavy metal-contaminated soils, including siderophores, ACC deaminase, and plant growth hormones(Ahmad et al., 2019). Because oxidative stress slows down the pace of phytoremediation, high soil contamination may hinder plant growth, particularly root growth and expansion. Phytoremediation is restricted by low microbial density and a shortage of nutrients. Manganese (Mn), cadmium (Cd), lead (Pb), chromium (Cr), zinc (Zn), aluminum (Al), and copper (Cu) are examples of common heavy metals. Arsenic (As) and antimony (Sb) are two examples of metalloids that exhibit toxicity. As microbes involved in phytoremediation, rhizosphere bacteria can directly increase process efficiency through changes in soil pH and oxidation/reduction reactions. Because mycorrhizal fungi are hyperaccumulators of heavy metals and have heavy metal tolerance, they are important components in phytoremediation. Current agricultural systems can be made more sustainable by utilizing beneficial bacteria. Plant nutrient control is greatly aided by members of the rhizosphere microbiome. Plants typically obtain their nutrients from the rhizosphere and the phyllosphere. Effective management of plant nutrients involves making the most use of soil, water, atmospheric conditions, and NPK fertilizers. A healthy microbiome can also help increase nutrient use efficiency. Many studies have been conducted on the benefits of symbionts like mycorrhizal fungi for growing plants. They were useful in improving soil structure and aggregates in crops like cereals, pulses, fruits, and oilseeds to meet their nutritional requirements. Several other bacterial endophytes, besides Rhizobium and Bradyrhizobium, have been documented to form symbiosis with plants for bioavailable nitrogen fixation in unspecialized host tissues, even when nodules are not present. For instance, heterocysts suitable for biological nitrogen fixation (BNF) with nitrogenase are formed by cyanobacteria in symbiotic associations with a variety of plants from diverse clads, including gunnera, cycads, and lichens.

6.12 RHIZOBIOMES

Microbiomes can help the plant body absorb a number of trace elements, including calcium and iron(Singh et al., 2022). Plant microbiomes break down organic substances that are necessary for plants to thrive in nutrient-poor soils as well as for their survival. Comprehending the interplay between microbes and plants is crucial for the cultivation of a favorable soil microbiome. It is uncertain if these advantageous microbial populations will persist in soil used for agriculture. In natural settings, the association of microbial communities with plant roots is primarily determined by two factors: the kind of soil and the roots of the plants. Plant root exudates and specific bacteria are directly impacted by the physiochemical features of soils. The soil biome is determined by the kind of soil, and plant root exudates tend to establish a rhizobiome that is favorable. To create a favorable rhizobiome and boost crop yields while lowering losses due to biotic and abiotic stresses, a variety of microbial communities are recruited by soil type and plant species, which together determine the composition of the rhizosphere. These elements play a major role in the selective enrichment of advantageous microorganisms in the rhizobiome, which may aid in identifying genetic features to enhance plant health and productivity.

Microbial communities have the potential to maximize crop production in terms of plant growth, yield, and resistance to biotic and abiotic stresses. The idea of using plant microbiomes to produce crops is not new. The plant microbiome has drawn a lot of attention recently and is a major factor in determining plant health and productivity. Improved resistance to biotic and abiotic stresses can be achieved through manipulating the plant microbiome.

This can also lead to increased agricultural productivity, a decrease in chemical inputs, and decreased greenhouse gas emissions. Maintaining the increasing world population depends on this. Agricultural practitioners are currently employing naturally occurring beneficial microorganisms to enhance crop plant performance. Even though the diversity and richness of species found in plant microbiomes are mostly unknown, the development of new plant hosts with a particular trait-associated microbiome is essential for the advancement of enhanced production systems. A rhizome that has the potential to significantly contribute to increased plant productivity in the future is supported by a variety of molecules, microbes, plant species, and mechanisms. Using both traditional and contemporary breeding techniques, several methods have been developed to modify root exudates to alter the rhizobiome and redirect microbial activity. There are currently efforts underway to duplicate or partially repair the plant microbiome/rhizobiome to create PGPB and/or PGPF consortia. Tomato plants infected with PGP consortia *(Bacillus amyloliquefaciens*, *Bacillus pumilus* T4, AMF *Glomus intraradices*) under greenhouse conditions produced a full yield with 30% fewer inputs. *Bacillus amyloliquefaciens* produces phytohormones that have been shown to increase *B. japonicum* nodulation efficiency. The benefits of microbial inoculants can be enhanced and stabilized by the genetic diversity of the local soil microbiome. As a result, it is advised that microbiome profiling be used in the field to determine, track, and applying microbial inoculants precisely where needed.

6.13 PLANT GROWTH

Plant growth is thought to benefit from arbuscular mycorrhizal fungi (AMF), which live in the rhizosphere and colonize plant roots(Amballa & Bhumi, 2016). By providing vital mineral nutrients, particularly phosphorus, AMF enhances the rhizosphere soil properties and benefits the host plants. It also prevents the transmission of destructive ions like Na and other metals. To protect themselves from the harmful effects of various environmental stresses, plants have developed many tolerance mechanisms. The accumulation of osmolytes and the selective absorption of ions boost the antioxidant system, which is the main mechanism of tolerance. Many studies on the potential of AMF in controlling plant development have been reported. The majority of researchers have focused on morphological alterations in the host plant and have hardly discussed the physiological, biochemical, and molecular processes underlying AMF-induced stress tolerance and growth promotion. To prevent the oxidative damage caused by excess ROS, plants have evolved a number of defense mechanisms to maintain the ideal concentration of ROS. An antioxidant defense system made up of both enzymatic and nonenzymatic components is one of these systems.

Superoxide dismutase or SOD, catalase (CAT), guaiacol peroxidase (POD), ascorbate peroxidase (APX), monodehydroascorbate reductase (MDAR), dehydroascorbate reductase (DAR), and glutathione reductase (GR) are some of the enzymatic components; nonenzymatic components include vitamins like glutathione, polyphenols, and ascorbic acid and tocopherol. To shield cells from oxidative damage brought on by stress, the antioxidant system eliminates surplus reactive oxygen species. In contrast to other tolerance mechanisms, antioxidant systems are thought to be one of the primary defense mechanisms for reducing the harmful effects of stress. The build-up of osmolytes like free proline, sugar alcohols, and amino acids, which results in the preservation of cellular structures and function, is another important factor in stress reduction. By aiding in the scavenging of ROS, osmolytes support the antioxidant system when it becomes ineffective. Osmoregulation in salt-stressed conditions is mediated by effective uptake and accumulation of ions, such as potassium, sodium, and chloride; in drought-stressed conditions, stress tolerance is mediated by accumulation of compatible solutes and ions along with the simultaneous sequestration and compartmentation of toxic ions into vacuoles or apoplasts.

6.14 MYCORRHIZAL INOCULATION: THE BENEFITS

To combat stress, plants attempt to minimize any potential negative effects of stress at the initial stages of stress exposure. For instance, morphological adaptations include leaf rolling, the development of heavy pubescence on leaves, leaf falling, and structural modifications. Plants have a variety of defense mechanisms in addition to their innate tolerance system to safeguard metabolism at the physiological and biochemical levels. Arbuscular mycorrhizal fungi or AMF and plant growth-promoting rhizobacteria (PGPR) are the two main microorganisms in rhizospheric soil that have been shown to have a beneficial role(Nadeem et al., 2014). The majority of plants can be directly colonized by AMF, microscopic filamentous fungus, in the root tissues or through the rhizosphere. To benefit the host plant, AMF typically develops branched filaments that are a few centimeters long. AMF have been shown to have the ability to enhance soil structure and plant growth in both normal and stressed settings. They form symbiotic interactions with a variety of plant species. For instance, AMF are inoculated and thought to function as vital bio-ameliorators in salt- and heavy metal-stress conditions, assisting plants in reducing stress-induced damage. AMF colonization increases plant vigor and growth. AMF colonization alters plant morphology, nutrition, and physiology. It has also been shown to increase a plant's resistance to biotic and abiotic stresses. To increase root access to water and nutrients, AMF alters the architecture of roots.

Mycorrhizal inoculation influences both the physiology and the morphology of the roots in host plants. Citrus plants colonized with AMF, for instance, had larger leaf areas, more phosphorus, and more photosynthesis than uncolonized plants. AMF colonization of the tomato root revealed various morphological structures formed by the appressorium. It has been demonstrated that AMF symbionts are connected to over 90% of plants and, by offering many advantages to plants, significantly aid in the management of plant growth. AMF association and its ecological significance in terrestrial and wetland plants have been described. Several AMF isolates have been classified into six distinct genera and an order known as Glomales, which is also known as Zygomycota. Ribosomal RNA sequencing used in molecular investigations verified that all AMF species belong to the same monophyletic group and are distinct from other fungus. The AMF isolates that infect roots or rhizospheres may now be directly identified using molecular techniques. Additionally, genetic diversity among the many species has been verified. An AMF spore can have thousands of nuclei on its surface, indicating its multigenomic nature. It should be mentioned that depending on the characteristics of the AMF or the host, an the genotype of the AMF to influence a host plant varies greatly.

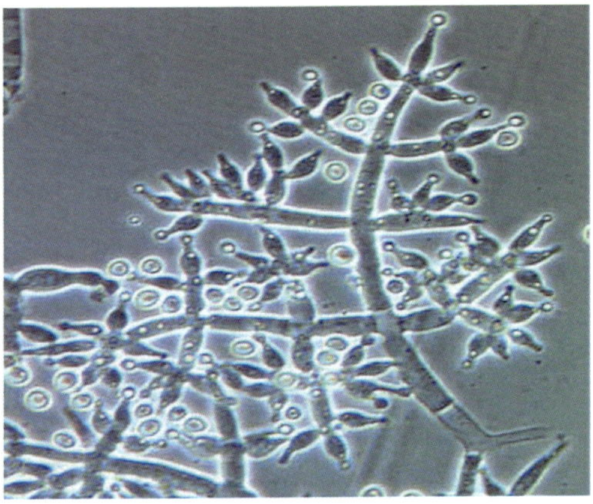

Figure 6.4*Trichoderma harzianum*

Source: https://upload.wikimedia.org/wikipedia/commons/4/4d/Trichoderma_harzianum.jpg,*Creative Commons License*

For instance, the absorption of mineral nutrients from the soil solution and their subsequent transfer into the host plant are carried out by the extra-radical mycelia of AMF. Even in the presence of AMF, a plant's capacity to assimilate minerals is impacted by the constitutive expression of transport proteins that are involved in the assimilation of critical elements like nitrogen and phosphorus. It is, therefore, necessary to acknowledge that differences exist in the ability of various AMF isolates or species to enhance plant development and safeguard plant metabolism and that these differences rely on the specific AMF isolate and host species. Water stress is a significant abiotic stressor that hinders several essential metabolic processes in plants, such as protein synthesis, photosynthesis, enzyme activity, nutrient uptake and assimilation, and antioxidant metabolism. It has been noted that temperature stress frequently coexists with water stress, causing the soil to dry up excessively and eventually leading to drought stress. Water stress is the lack of available water for roots to absorb; on the other hand, higher temperatures accelerate transpiration, which coincides with photosynthetic arrest brought on by excessive production and buildup of reactive oxygen species (ROS), a type of oxidative stress.

Trichoderma harzianum(Figure 6.4) enhances soil properties that lead to increased yield, chlorophyll production, and nitrogen uptake, hence protecting Triticum aestivum from drought stress(Shukla et al., 2015). Glomalin formation and hyphal network intrusion may contribute to increased uptake of water and nutrients and subsequent growth promotion following AMF inoculation. These processes enhance soil structure and consequently water interactions. Soil stabilization may also contribute to the higher plant growth that AMF mediates. Many researchers have linked the AMF-induced enhanced growth and reduction of drought stress to improvements in the uptake and transfer of nutrients and water, which improve osmoregulation and water-use efficiency and decrease ROS generation. Improvements in these characteristics may be attributable to increased water uptake in AMF-inoculated plants, which is caused by the presence of extraradical hyphae. AMF inoculation protects photosynthesis and related attributes like stomatal conductance and gas exchange, transpiration, and water-use efficiency. By raising the osmotic buildup and hydraulic conductivity, AMF inoculation promotes development during drought stress. Some studies showed overexpression of the root aquaporin gene, which is important in water uptake in Zea mays, reports better water absorption in AMF-inoculated plants; this expression was specific in regions that were infected by hyphae. In comparison to uninoculated plants, cucumber plants inoculated with *Trichoderma harzianum* exhibited increased uptake of critical elements like N, P, K, and Cu. This suggests that soil-borne microorganisms may be actively involved in directly modulating nutrient uptake in addition to enhancing soil health. By preserving the tissue water potential, stimulation of the production of osmolytes like proline, glycine, betaine, sugars, and polyols might enhance the growth-promoting effect of AMF in plants. When AMF is inoculated into Erythrina variegata, the amount of sugar production is reduced in comparison to control plants. Similar to this, when plants were exposed to drought stress, AMF-inoculated plants displayed variable amino acid synthesis. The accumulated production of amino acids caused by AMF can safeguard cellular function by preserving the precursor pools of many essential compounds, including redox components like glutathione. Many studies support the involvement of osmolytes in the ability to withstand water stress; yet, the function of AMF and other soil-borne microflora is still unclear and needs further research. As a result of proline's critical role in stress tolerance, proline synthesis is downregulated in samples inoculated with AMF.

Certain plant species have evolved defense mechanisms like volatilization to eliminate heavy metals from their bodies when they acquire significant quantities of them. These plants are good candidates for phytoremediation because they can sustain the abundant growth of symbionts (PGPR or AMF), which increases their resistance to harmful metals. Soil-borne microflora alters the oxidation state of hazardous metals, either decreasing their toxicity or mediating their removal as complexes. Similar to this, AMF increases plant development by altering the pH of rhizospheric soil and chelating or precipitating metals in chemicals secreted by related fungal species, such as polyphosphate granules. Hazardous metals like copper and cadmium are prevented from moving by the glomalin released by AMF. Even in highly saline environments, AMFs are found in nearly all types of soil. Significant variability has been observed in the behavior of AMF; for example, in certain situations, it enhances the transfer of heavy metals to host plants while preventing their mobilization in other situations, giving rise to two distinct processes known as phyto-extraction and phyto-stabilization. The gene (*Gigaspora margarita*) that codes for the metallothionein decrease metal toxicity(Pócsi, 2011). However, there are examples of hazardous metal chelation in the rhizosphere. Plants colonized by AMF have comparatively lower levels of harmful metal ion buildup/accumulation in their shoot tissue. In AMF-inoculated pea seedlings exposed to arsenic stress, researchers found higher leaf water content and chlorophyll concentrations; these effects were caused by the cumulative effect of AMF on the accumulation of proline, glycine betaine, and total free amino acids.

Chapter 7

METABOLISM AND PHYSIOLOGY

Contents

7.1 INTRODUCTION

The metabolism of organic substances by bacteria other than glucose is very crucial. It can be presumed that an organism can use a substance if it is converted to intermediates of central metabolism because central metabolism is reversible. Certain organisms can only use a restricted number of organic compounds; *Bacillus fastidiosus*, for instance, can only use urate(Izaguirre-Mayoral et al., 2018). Other bacteria, on the other hand, can use a range of organic compounds as their primary sources of carbon and energy. The primary component of plant and animal cells is polymer. They consist of proteins, nucleic acids, and other substances in addition to carbohydrates like cellulose and starch. Such polymers cannot be easily carried into microbial cells but are first degraded to monomers or oligomers by extracellular enzymes before being transferred inside the cell.

7.2 PRODUCTION

The glucose polymer known as starch is made up of amylose and amylopectin. The chain structure of the former is straight with α-1,4-glucoside linkages, whereas the side chains of the latter contain α-1,6-glucoside bonds. Since starch is an abundant storage substance in plants, a lot of prokaryotes utilize amylase as a source of carbon and energy. Amylases are categorized based on how they work. The endoglucanase α-amylase hydrolyzes at random α-1,4-glucoside binds to create a mixture of glucose, maltose, and dextrin; however, it does not hydrolyze α-1,6 bonds. This enzyme is produced by several bacteria, including *Bacillus*, *Pseudomonas*, and *Clostridium* species. Maltose units are extracted from the non-reducing end of amylose by the exoglucanase, a α-amylase. This enzyme is produced by many bacteria, such as Pseudomonas species and Bacillus. Fungi are used in the industrial production of glucoamylase. Amylases break down the α-1,4-glucoside bond of amylose, yet it is unable to hydrolyze the α-1,6-glucoside bond.

Pullulanase and isoamylase hydrolyze the α-1,6-glucoside bond in amylopectin. The fungus *Aureobasidium pullulans* produces pullulan, an α-glucan, which is hydrolyzed by pullulanase and attacked by isoamylase. α-1,6-glucoside links glycogen and amylopectin. We call these enzymes debranching enzymes. Pullulanase is produced by *Aerobacter aerogenes*, *Bacillus cereus*, and other bacteria; isoamylase is produced by *Bacillus amylo liquefaciens*. An enzyme that can hydrolyze α-1,4/ α-1,6-glucoside bonds and is produced by a few hyperthermophilic bacteria and archaea, such as *Thermoanaerobacter ethanolicus*, *Pyrococcus furiosus*, and *Thermococcus litoralis*. This highly thermostable enzyme is known as an amylopullulanase or neopullulanase. Similar to starch, cellulose is a glucose homopolymer and is one of the most prevalent organic substances. The cellulose is more resistant to hydrolysis than starch. In contrast to starch, cellulose is a straight chain of β-1,4-linked glucose units without any side chains. Cellulose produces a crystalline structure through hydrogen bonding between its molecules. The hydrolysis of crystalline cellulose requires the involvement of some enzymes. They are cellobiase or β-glucosidase, endo-β-glucanase, and exo-β-glucanase. These enzymes are produced by certain bacteria, including *Clostridium thermocellum* and *Cellulomonas flavigena*. These combine with other proteins to create a new cellulosome (complex). It is unknown if the several endo-β-glucanases and exo-β-glucanases produced by these bacteria are the result of distinct genes. Certain bacteria are limited to using amorphous cellulose.

The enzyme β-glucan cellobiohydrolase which breaks down crystalline cellulose is absent. β-1,4-glucoside bonds which are buried inside the crystalline structure, are hydrolyzed by cellulases. Several theories explain how the enzyme hydrolyzes crystalline cellulose because the enzymes are too large to make direct contact with the hydrolytic sites.

A possible explanation is that endo-β-1,4-glucanase hydrolyzes the amorphous region, producing a lot of non-reducing ends that β-glucan cellobiohydrolase, an enzyme belonging to the exo-β-1,4-glucanase enzyme class, can use to remove cellobiose units. The hydrolysis of crystalline cellulose is accomplished by the coordinated action of these enzymes. Cellobiase either transports cellobiose into the cell or hydrolyzes it to glucose.

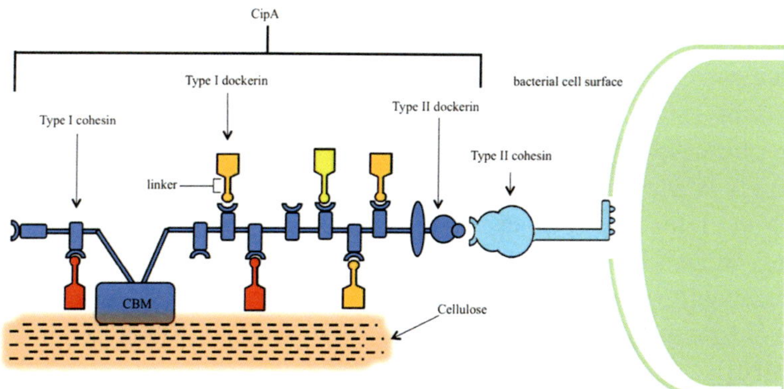

Figure 7.1 *Clostridium thermocellum*

Source: Akinosho H, Yee K, Close D and Ragauskas A (2014) The emergence of *Clostridium thermocellum* as a high utility candidate for consolidated bioprocessing applications. Front. Chem. 2:66. doi: 10.3389/fchem.2014.00066, *Creative Commons License*

7.3 ENZYMES SOURCED FROM MICROBES

Before the hydrolytic enzymes target the β-glucose bonds, a C1 enzyme recrystallizes the native cellulose to the amorphous form with negligible hydrolytic activity(Chen, 2014). X-ray crystallography has supported this observation. Strong cellulose-binding activity has been observed in an enzyme present in fungi like *Trichoderma reesei* and bacteria like *Clostridium thermocellum*(Figure 7.1), *Clostridium cellulovorans*, and *Cellulomonas flavigena*. Other naturally occurring polysaccharides include hemicellulose, pectin, and chitin in addition to starch and cellulose. A heteropolysaccharide known as hemicellulose is made up of several derivatives of pentoses and hexoses that are joined together by β-glycoside bonds. Using the extracellular enzymes known as hemicellulase, a variety of bacteria, including cellulolytic species, can exploit this heteropolysaccharide as a carbon and energy source. Most hemicelluloses contain xylose as the predominant monosaccharide; hemicellulase is sometimes called xylanase. Pectin is an α-1,4-polygalacturonate methyl ester. Endo- and exotype pectinases break down the ester link, whereas pectin esterase hydrolyzes the ester bond to create methanol. These enzymes can be produced by different bacteria, including *Erwinia carotovora* and *Bacillus polymyxa*. These enzymes can be used commercially to clarify fruit juices. Chitin (Fungi) and the exoskeletons of insects and crustaceans have the structure of poly-β-1,4-N-acetylglucosamine. After cellulose, this polymer is the second most prevalent in the biosphere. Because of its crystalline form, chitin needs many enzymes to break down. *Chromobacterium violaceum*, *Serratia marcescens, Serratia plymuthica*, *Serratia liquefaciens*, *Aeromonas hydrophila*, *Enterobacter agglomerans*, *Pseudomonas aeruginosa*, *Pseudomonas chitinovorans*, and *Streptomyces griseus* are just a few of the soil bacteria that produce chitinase. The enzyme creates N-acetylglucosamine by hydrolyzing β-1,4 bonds. Acetate is extracted from chitin by the enzyme chitin deacetylase, resulting in chitosan, which chitinase then hydrolyzes to glucosamine. Fungi including *Absidia cierulea*, *Colletotrichum lindemuthanum*, and *Mucor rouxii* have chitin deacetylase. Chitinase-producing microbes can function as biocontrol agents by preventing the growth of plant pathogenic fungi (Neeraja et al., 2010).

Nature contains a variety of different polysaccharides. Extracellular enzymes, which are produced by a variety of microbes, also hydrolyze these. Enzymes that hydrolyze polysaccharides are categorized based on the order of their amino acids. Non-hydrolytic enzymes allow the organism to import and use some disaccharides. Using inorganic phosphate, they phosphorylate and release monosaccharide in its free form. Phosphorylase is the term used to describe these enzymes. Phosphorylases use less ATP than hydrolases in the metabolism of disaccharides because hexokinase is not engaged in the metabolism of the phosphorylated sugar. *Pseudomonas saccharophila* contains sucrose phosphorylase, while *Cellulomonas flavigena* phosphorylates cellobiose. The utilization of environmental polymers as sources of energy and carbon is not the function of these enzymes. Extracellular enzymes are produced by bacteria that use extracellular polymers. Based on their ideal pH, extracellular proteases are categorized as acidic, neutral, or alkaline enzymes. *Bacillus licheniformis* produces the alkaline protease subtilisin, which has an ideal pH range of 8 to 11. *Pseudomonas aeruginosa* and *Bacillus megaterium* produce neutral proteases. The majority of alkaline proteases hydrolyze all peptide bonds and are non-specific. Many distinct acidic proteases hydrolyze the peptide bonds of specific amino acids. The carbon and energy sources for bacteria that generate extracellular DNase and RNase, respectively, are DNA and RNA. Extracellular DNase is produced by *Staphylococcus aureus*, while RNase is known to be present in *Bacillus subtilis*, Streptomyces species, and other bacteria. Microorganisms use lipids that extracellular lipase has hydrolyzed into fatty acids and glycerol.

7.4 METABOLIC INTERACTIONS

Several hexoses, pentoses, and their derivatives are produced by the hydrolysis of extracellular polysaccharides. These materials facilitate the growth of several bacteria and archaea. Hexokinase phosphorylates fructose to fructose-6-phosphate following active transport into the cell, or fructose-1-phosphate during translocation. 1-phosphofructokinase then phosphorylates fructose-1-phosphate to fructose-1,6-diphosphate.

Prior to being converted to mannose-6-phosphate by hexokinase, mannose is actively carried into the cell. Mannose-6-phosphate is isomerized to fructose-6-phosphate by phosphomannoisomerase. To create galactose-1-phosphate and glucose: galactose-1-phosphate, galactokinase transfers phosphate from ATP to galactose. UDP-galactose and glucose-1-phosphate are created when uridylyltransferase transfers UMP from UDP-glucose to galactose-1-phosphate. UDP-glucose epimerase is responsible for converting UDP-galactose to UDP-glucose. Glucose-1-phosphate is changed into glucose-6-phosphate by phosphoglucomutase. Glucose-1-phosphate facilitates the metabolism of glycogen. Specific phosphorylases or hydrolases such as α-glucosidase, invertase, or β-glucosidase (cellobiase) cleave the disaccharides. Before β-galactosidase hydrolyzes lactose to produce glucose and galactose, lactose is actively transported. Common pentoses include ribose, arabinose, and xylose. Phosphorylation of glucose results in ribose-5-phosphate. Isomerases change arabinose and xylose into ribulose and xylulose, respectively. The HMP pathway is used to detoxify these ketoses after they have been phosphorylated. Before acyl-CoA is transformed into acetyl-CoA by β-oxidation, acyl-CoA synthetase synthesizes acyl-CoA from fatty acids and coenzyme-A. Acetyl-CoA is converted into the carbon skeletons required for biosynthesis by the TCA cycle in conjunction with the glyoxylate cycle. β-oxidation, which is mediated by many enzymes, including acyl-CoA synthetase, splits a 2-carbon unit from acyl-CoA into acetyl-CoA. When fatty acids with an odd number of carbons are oxidized, propionyl-CoA remains, but fatty acids with an even number of carbons only produce acetyl-CoA. One of the metabolic intermediates such as the amino acids L-valine and L-isoleucine, is propionyl-CoA. Research revealed the acrylate pathway and the methylmalonyl-CoA pathway in the oxidative metabolism of propionyl-CoA. Researchers have demonstrated that the methyl citrate cycle is the mechanism by which propionyl-CoA is converted to pyruvate in a variety of bacteria, including *Escherichia coli*, *Salmonella typhimurium*, and *Ralstonia eutropha*, as well as in fungi, including *Saccharomyces cerevisiae* and *Aspergillus nidulans*. Propionate can prevent fungal growth, but it can also act as a source of energy and carbon.

Pseudomonas fluorescens and *Acinetobacter calcoaceticus* activate malonate, a succinate dehydrogenase inhibitor, to malonyl-CoA, which is then decarboxylated to acetyl-CoA. Certain bacteria, like the anaerobe *Malonomonas rubra* and the bacterium *Klebsiella pneumoniae*, use an acyl carrier protein (ACP) in place of coenzyme A when they metabolize malonates(Dimroth & Hilbi, 1997). For the TCA and glyoxylate cycles, acetate can be the only carbon source. The metabolism of C2 compounds differs from that of acetate because they are more oxidized. These substances include oxalate, glycolate, and glyoxylate. Phosphoglycolate, which is formed during photorespiration, is dephosphorylated to produce glycolate. The breakdown of purines produces glyoxylate. These materials are utilized by Escherichia coli and Pseudomonas species via the glycerate pathway to furnish carbon skeletons for biosynthesis and the dicarboxylic acid cycle to produce energy. The glycerate pathway replenishes phosphoenolpyruvate, which is needed for biosynthesis and is an intermediary of the dicarboxylic acid cycle. Glyoxylate is converted to oxaloacetate, a TCA cycle intermediate, by *Paracoccus denitrificans* via the 3-hydroxyaspartate pathway. Glyoxylate is changed into glycine in the 3-hydroxyaspartate pathway, and then it condenses with another glyoxylate molecule to produce erythro-3-hydroxyaspartate. This is converted to oxaloacetate by erythro-3-hydroxyaspartate dehydratase. This enzyme is called a dehydratase even though it deaminates the substrate, just like in the deamination of amino acids with a hydroxyl group. Oxalate serves as the only carbon and energy source for Pseudomonas oxalaticus. This oxidized dicarboxylic acid is reduced to glyoxylate for biosynthesis and then activated to oxalyl-CoA. A portion of it is then oxidized to CO_2 as an energy source. Both *Methylobacterium extorquens* and *Bacillus oxalophilus* rely only on oxalate as their energy and carbon source(Cheng et al., 2023). Butanol and other alcohols are produced by anaerobic bacteria. Primary alcohols are converted to fatty acids by the enzymes alcohol dehydrogenase and aldehyde dehydrogenase. Before secondary alcohol is converted to a fatty acid and primary alcohol by the action of a monooxygenase and esterase, secondary alcohol dehydrogenase oxidizes secondary alcohol to a ketone.

7.5 SUBSTRATES AND INTERMEDIATE METABOLISM

Glycerol ferments to produce propanediol, while some Bacillus species and facultative anaerobic bacteria use glucose as a source of butanediol and acetoin. Acetone and isopropanol are two of the fermentation products produced by saccharolytic clostridia from carbohydrates. Propanediol is oxidized by *Salmonella typhymurium* to propionyl-CoA, which is then metabolized via the methylcitrate cycle and other routes. Through the actions of diol dehydrogenase and the acetoin dehydrogenase complex, species of Bacillus, *Ralstonia eutropha*, and *Pelobacter carbinolicus* oxidize butanediol to acetaldehyde and acetyl-CoA. Similar to the pyruvate dehydrogenase complex, the acetoin dehydrogenase enzyme complex is a keto acid dehydrogenase. Isopropanol can be converted by *Xanthobacter* sp. to acetone and then carboxylated to acetoacetate.

Thiosphaera pantotropha, *Rhodomicrobium vannielii*, and *Rhodobacter capsulatus* carboxylate acetone. Nutrient broth, which includes peptone and beef extract, is a popular medium for growing a wide variety of bacteria in the lab. These compounds are made up of peptides and amino acids. Amino acids and peptides are transported and metabolized by organisms living on such nutritional media via central metabolic pathways. Protein synthesis requires amino acids, which are deaminated to produce 2-keto acids. 2-keto acid dehydrogenases oxidize the 2-keto acids to acyl-CoA, which can be used as a source of carbon and energy. Amino acid dehydrogenase reduces NAD(P). A single enzyme can oxidize up to ten distinct amino acids, and amino acid oxidases have a low substrate selectivity. Bacteria have both L-amino acid and D-amino acid oxidase because their cell walls contain D-amino acids(Takahashi et al., 2015). L-alanine or L-glutamate is oxidized by amino acid dehydrogenase to produce 2-ketoglutarate and pyruvate, respectively. The combination of transaminase and amino acid dehydrogenase can deaminate all amino acids since transaminases convert pyruvate and 2-ketoglutarate to alanine and glutamate. An enzymatic process called transamination $-NH_2$ group to 2-keto acids. Alanine and glutamate dehydrogenases deaminate substrate. Combining glutamate or alanine transamination with dehydrogenation results in the oxidation of an amino acid to the equivalent 2-keto acid, which lowers $NAD(P)^+$.

Amino acid dehydratases simultaneously remove the hydroxyl group (-OH) and deaminate serine and threonine. Threonine dehydratase catalyzes the deamination of threonine to 2-ketobutyrate, which is the first step in the biosynthesis of isoleucine. For the production of isoleucine and the utilization of threonine as a source of energy and carbon, *Escherichia coli* possesses distinct threonine dehydratases(Kagan et al., 1975). Aspartase and histidase create double bonds between two and three carbons, in contrast to dehydrogenases and oxidases. Neither water nor electron carriers participate in these processes. Homocysteine is produced when transmethylase extracts the methyl group from methionine. Pyruvate and 2-ketobutyrate are the products of desulfhydrase removal of amino and sulfide groups from homocysteine and cysteine, respectively. Many facultative anaerobic and aerobic bacteria, such as *Bacillus subtilis*, *Proteus vulgaris*, and *Escherichia coli*, are known to have desulfhydrase. The central metabolic intermediates pyruvate, oxaloacetate, fumarate, and 2-ketoglutarate have both anabolic and catabolic uses.

Depending on the organism, glyoxylate is metabolized via either the 3-hydroxyaspartate pathway or the dicarboxylic acid cycle–glycerate pathway. 2-Ketobutyrate is oxidized by a dehydrogenase complex to propionyl-CoA, which is then metabolized via the methylcitrate cycle and other pathways. When an amino acid has a side chain, it can oxidize to produce 3-methylvalerate, 2-ketoisovalerate, and 2-ketoisocaproate. The 2-keto acid dehydrogenase complex further oxidizes them to acyl-CoA before acyl-CoA dehydrogenase produces a double bond between two and three carbons. All of these reactions are catalyzed by the same enzyme. A different enzyme breaks down unsaturated fatty acids with side chains into propionyl-CoA, acetyl-CoA, and acetoacetate. Two acetyl-CoA are produced via acetoacetate and acetoacetyl-CoA.

Depending on the organism, arginine can be converted via a variety of distinct mechanisms to either succinate or glutamate. Through processes mediated by proline dehydrogenase, which also possesses proline oxidase and 1-pyrroline-5-carboxylate (P5C) dehydrogenase activity, enteric bacteria convert proline to glutamate. The initial proline oxidation process results in the reduction of FAD, then the same enzyme reduces NAD+ by oxidizing P5C to glutamate. When proline is available to act as an electron donor, this enzyme binds to the cytoplasmic membrane; however, when proline is not present, the Put (proline dehydrogenase and proline permease) operon serves as the binding site of the enzyme on the chromosome, inhibiting transcription of the operon.

A sequence of enzymatic events converts lysine to 2-keto-6-aminocaproate by a transamination reaction, which is then followed by degradation to acetyl-CoA via 2-aminoadipate, glutaryl-CoA, and acetoacetyl-CoA. Of all the amino acids, tryptophan breakdown is the most intricate. The processes are started by tryptophan-2,3-dioxygenase, which opens the pyrrole ring to produce formylkynurenine. Kynurenine then breaks this down to produce alanine and anthranilate. DNA and RNA are hydrolyzed by nucleases to produce deoxyribonucleoside monophosphate and ribonucleoside monophosphate. Nucleotidase and nucleosidase hydrolyze them further to bases and ribose (deoxyribose). While adenine requires two processes to be oxidized to xanthine via hypoxanthine, adenine deaminase and xanthine dehydrogenase (xanthine oxidase) catalyze the process, guanine deaminase catalyzes the one-step reaction from adenine to xanthine(Pang et al., 2012). Depending on the organism, either xanthine oxidase or xanthine dehydrogenase oxidizes hypoxanthine and xanthine. In addition to oxidizing hypoxanthine to xanthine, xanthine dehydrogenase or oxidase also converts the subsequent reaction product to urate. Superoxide is produced when xanthine oxidase reduces molecular oxygen, whereas xanthine dehydrogenase employs NAD(P)+ or ferredoxin as its electron acceptor. A sequence of events converts urea and glyoxylate from ureate. Glyoxylate is metabolized via the hydroxyaspartate pathway or the dicarboxylic acid cycle–glycerate pathway.

7.6 HYDROCARBONS

Aliphatic hydrocarbons are used by a wide variety of prokaryotic and eukaryotic microorganisms, particularly those that are liquid at room temperature. Hydrocarbons are different from water-soluble substrates since they are water-insoluble. At the water–oil interface, microorganisms use hydrocarbons as a source of carbon and energy because they cannot survive in a pure oil phase. The cell surfaces of bacteria and fungi that consume hydrocarbons include hydrophobic glycolipids. Before being delivered into the cell, the hydrocarbon is made soluble by this glycolipid. *Acinetobacter calcoaceticus* is one of many bacteria that create a surfactant to enhance hydrocarbon transport in emulsions. At the cytoplasmic membrane, hydrocarbon monooxygenase converts a hydrocarbon to a primary alcohol. This enzyme uses a single oxygen molecule to oxidize two distinct substrates. This class of enzyme is known as hydroxylase. Alcohol dehydrogenase and aldehyde dehydrogenase catalyze processes that further oxidize the alcohol formed from the hydrocarbon oxidation to a fatty acid. *Nocardia petroleophilia* monooxygenase oxidizes the second carbon (hydrocarbon) to create a secondary alcohol, which is then further oxidized to a ketone by a secondary alcohol dehydrogenase. The ketone is converted to a primary alcohol and acetate by a second monooxygenase and acetylesterase. Propylene is converted to acetoacetate by *Rhodococcus rhodochrous*. These 3-carbon compound intermediates are metabolized and bound to coenzyme M (2-mercaptoethanesulfonic acid), a C1 carrier in methanogenic archaea. Coal and petroleum contain a range of aromatic chemicals.

Because of their structural stability, these materials oxidize mostly in anaerobic environments. Aliphatic hydrocarbons are readily oxidized. The breakdown of aromatic hydrocarbons is mostly recognized in Pseudomonas species, such as *Pseudomonas acidovorans* and *Pseudomonas putida*. Phenylalanine is oxidized to tyrosine by phenylalanine monooxygenase, which is then deaminated to produce p-hydroxyphenylpyruvate. Dioxygenases or oxidases are enzymes that combine two atoms of oxygen into a single substrate. When the benzene ring is opened by Homogentisate oxidase, fumarate and acetoacetate are eventually produced. Monooxygenases add hydroxyl groups to the benzene ring in the first stage. Three intermediates are produced from aromatic chemicals in this step: protocatechuate, catechol, and gentisate. Aromatic compounds containing amino groups, and lignin monomers are the main sources of protocatechuate, which is the end product of protocatechuate conversion.

7.7 DIOXYGENASES

Dioxygenases release these intermediate benzene rings(Wang et al., 2017). Dioxygenases are responsible for catalyzing ring fission, which is known as ortho cleavage when it takes place in the space between hydroxyl groups and meta cleavage when it happens next to one of the hydroxyls. The oxidation of aromatic compounds containing hydroxyl groups at para locations is caused by the gentisate pathway, another ring cleavage process. Following the ring fission processes, succinate, acetyl-CoA, pyruvate, and acetaldehyde are the end products that are metabolized. The genes for the ortho cleavage pathway are located in the chromosomes of bacteria, whereas the genes for the meta cleavage pathway are located on plasmids. The term degradative plasmid is used to distinguish these plasmids from others that serve various purposes, like resistance to antibiotics. The gentisate pathway, which is initiated by gentisate dioxygenase, breaks down gentisate and its derivatives to pyruvate and fumarate. Maleylpyruvate is either immediately degraded to fumarylpyruvate or isomerized to it, depending on the organism. Since oxygenase in aerobic organisms initiates hydrocarbon degradation, it was thought that molecular oxygen was necessary for the degradation of aromatic chemicals. Nonetheless, many investigations have demonstrated that anaerobic environments can cause aromatic compounds to oxidize.

7.8 METHANE-METABOLIZING MICROBES

Large amounts of methanol, methane, and methylamines are found in nature. It is known that certain bacteria and yeasts utilize them as their only sources of energy and carbon. Methylotrophs metabolize C1 molecules via unidentified mechanisms in the metabolism of multicarbon compounds. The term methylotroph refers to all species that use C1, while in other sense, it describes co-organisms that use C1 but are unable to use methane. Based on their capacity to consume methane, methylotrophs are classified as either methylotrophs or methanotrophs. C1 is used by methanotrophs. Catechol is the metabolic pathway for aromatic chemicals. Catechol is the metabolic intermediary of lignin monomers, aromatic hydrocarbons, and aromatic compounds containing amine groups; multicarbon compounds are not used in this process. We refer to these as obligate methylotrophs. Methane is not used by methylotrophs. Methylotrophs are classified as either autotrophic or heterotrophic based on how they metabolize carbon. Eukaryotes do not have methanotrophs, and methylotrophic yeasts use the xylulose monophosphate pathway to absorb formaldehyde.

Methanotrophs are unable to exploit multicarbon molecules and must obtain their carbon and energy from C1 compounds. They are known as obligate methylotrophs. They employ C1 chemicals, but they also possess a few other traits. Like the nitrifiers, they are all gram-negative and contain large intracellular membrane structures. Methanotrophs are categorized using this membrane structure. Except for Methylobacterium species, all obligate methylotrophs are known to exhibit spore and cyst forms during their resting state. Methane monooxygenase uses NADH as a cosubstrate to oxidize methane to methanol(Bennett et al., 2018). The electron transport system for ATP production receives electrons from the oxidation of methanol. Carbon dioxide or formaldehyde is absorbed for the production of cell components. Pyrroloquinoline quinone (PQQ) is reduced by methanol dehydrogenase in tandem with the oxidation of methanol to formaldehyde. Apart from methanol dehydrogenase, PQQ functions as a cofactor for alcohol dehydrogenase in acetic acid bacteria and glucose dehydrogenase in *Pseudomonas putida*. Either free or bound forms of carbon dioxide are produced by the oxidation of formaldehyde. Formaldehyde dehydrogenase and formate dehydrogenase, found in Gram-positive methylotrophs like *Amycolatopsis methanolica* and the autotrophic methylotroph *Paracoccus denitrificans*, oxidize formaldehyde in the free form to CO_2 via formate. In many Gram-negative bacteria that convert formaldehyde to cell materials via the ribulose monophosphate or serine–isocitrate lyase pathway, formaldehyde is linked to tetrahydrofolate (H4F) or to tetrahydromethanopterin (H4MTP) before being oxidized to CO_2. H4MTP is considered a coenzyme specific to methanogens. The enzymes found in *Methylobacterium extorquens* catalyze the oxidation of C1 molecules attached to either H4F or H4MTP. It is thought that formaldehyde is oxidized in this bacterium after being attached to H4MTP since the enzyme activities for C1 compounds bound to H4MTP are higher. There are no enzymes in *Methylophilus methylotrophus* that can oxidize formaldehyde either in its free form or when it is attached to C1 carriers. Through the ribulose monophosphate cycle, this bacterium converts formaldehyde to CO_2. Multicarbon compounds are inaccessible to obligate methylotrophs as a carbon source. Both heterotrophic and obligatory methylotrophs use the serine-isocitrate lyase (SIL) pathway to assimilate formaldehyde and CO_2, or the ribulose monophosphate (RMP) pathway to assimilate formaldehyde. CO_2 is fixed by autotrophic methylotrophs using the Calvin cycle. The xylulose monophosphate (XMP) route is another pathway used by methylotrophic yeasts that grow on methanol.

7.9 FORMALDEHYDE

Formaldehyde is converted into triose phosphate by the RMP route, which consists of distinct pathways that share some processes. Hexulose-6-phosphate is formed when RMP absorbs formaldehyde and is subsequently isomerized to fructose-6-phosphate. Two different methods are used to cleave fructose-6-phosphate: 2-keto-3-deoxy-6-phosphogluconate aldolase and fructose-1,6-diphosphate aldolase. Assimilatory metabolism uses the triose phosphate and pyruvate that are produced. All of the carbons in triose phosphate produced via the RMP route come from formaldehyde; however, *Methylosinus trichosporium* uses formaldehyde along with CO_2 to produce 2-phosphoglycerate through the acetyl-CoA pathway(Hou, 1984). There are no known methane absorbers in eukaryotes, and only a few yeast strains, including species of Candida, Hansenula, and Torulopsis, are known to utilize methanol. Methanol oxidase catalyzes the oxidation of methanol to produce hydrogen peroxide. In these yeasts, an organelle called the peroxisome has the enzymes catalase and methanol oxidase. Microorganisms use the Calvin cycle, SIL, XMP, and RMP pathways to ingest C1 molecules. The RMP pathway EMP-TA version is the most effective, whereas the Calvin cycle is the least effective. Part of the substrate is transformed into cell components during heterotrophic metabolism in aerobic environments, while the remaining portion is oxidized to produce carbon dioxide, which provides energy for growth. Because the conditions are such that the enzymes are inhibited or because there are insufficient enzymes for full oxidation, some microorganisms release massive amounts of metabolic intermediates. The synthesis of acetate by acetic acid bacteria and acetoin and butanediol by Bacillus species members are examples. Many alcohols and ketones are oxidized by acetic acid bacteria. To produce ascorbic acid, this characteristic is used to oxidize D-sorbitol to L-sorbose. Bacillus species that grow on carbohydrates, such as *Bacillus polymyxa* and *Bacillus subtilis*, have inhibited TCA cycle enzyme production. During cell growth, they do not fully oxidize the substrate, collecting acetoin and 2,3-butanediol. This metabolism bears similarities to facultative anaerobic enteric bacterial fermentation. Using acetoin and 2,3-butanediol as energy sources through the TCA cycle, Bacillus species sporulate when their supply of carbohydrates is depleted. A substance having two functional groups used as a raw material for polymer synthesis is 2,3-butanediol. Under anaerobic, nutrient-rich conditions, coryneform bacteria like *Brevibacterium flavum* and *Corynebacterium glutamicum* excrete enormous amounts of glutamate. The TCA cycle enzyme activity is minimal. Industrial use of this results in the production of monosodium glutamate. Certain ecosystems maintain anaerobic conditions when the rate of oxygen use exceeds the rate of oxygen supply. Other mechanisms, such as the events mediated by fumarate reductase and Na+-dependent decarboxylase, and lactate/H+ symport also contribute to the generation of ATP during fermentation. In some fermentations, like the ethanol and homolactate fermentation, the ratio of product created to substrate consumed is constant; in other fermentations, like the clostridial fermentation, it is variable.

Desulfovibrio gigas, *Desulfovibrio vulgaris*, and other Desulfovibrio species exhibit SOD and catalase activity, and they are less susceptible to O_2 than methanogens. Species of Desulfovibrio include proteins that detoxify reactive oxygen species, including superoxide, in addition to SOD and catalase. These electron carriers, which include rubredoxin, desulfoferredoxin, neelaredoxin, and rubrerythrin, contain iron. Neelaredoxin is a protein that serves as a SOD in the archaeon *Archaeoglobus fulgidus*, while superoxide reductase in *Pyrococcus furosus* breaks down superoxide directly into water(Dillard et al., 2011). Despite having SOD and catalase activity, certain strict anaerobes cannot grow in the laboratory in an aerobic environment. The presence of dissolved oxygen elevates the redox potential, and a high redox potential prevents the growth of some strict anaerobes. Methanogens have a redox potential of less than -0.3 V. Certain enzymes require sulfide, which is oxidized by molecular oxygen to produce disulfide. It is possible that the organisms cannot develop with such inactive enzymes. It is highly probable that the suppression of development by oxygen is caused by a combination of processes.

7.10 FERMENTATION PATHWAYS

Saccharomyces cerevisiae uses the ED pathway to ferment carbohydrates into ethanol, while Zymomonas mobilis uses the EMP pathway. Acetaldehyde is the product of decarboxylation of pyruvate and is utilized as the electron acceptor. When acetaldehyde is converted to ethanol, electrons produced during the glycolytic process are consumed. *Saccharomyces cerevisiae* produces two ATP from a single hexose molecule, whereas *Zymomonas mobilis* produces one ATP from a single hexose molecule. Just like pyruvate dehydrogenase, pyruvate decarboxylase also contains thiamine pyrophosphate as a prosthetic group. This enzyme is also present in *Sarcina ventriculi*, a strictly anaerobic acidophile, and *Erwinia amylovora*, a facultative anaerobe, in addition to Zymomonas mobilis. One important enzyme in the production of ethanol is pyruvate decarboxylase. It may be mentioned that enteric bacteria, heterofermentative lactic acid bacteria, and saccharolytic clostridia all create ethanol via distinct processes. Before converting pyruvate to ethanol, these bacteria oxidize it to acetyl-CoA.

Ethanol fermentation by acetyl-CoA is a branched process that yields a variety of fermentation products like lactate, acetate, and H_2, ethanol fermentation via pyruvate decarboxylase is a linear system that only produces CO_2 and water as byproducts. Thermophilic anaerobes use acetyl-CoA to ferment a variety of carbohydrates, such as cellulose and pentoses, into ethanol. *Thermoanaerobacter ethanolicus*, *Thermoanaerobium brockii*, and *Clostridium thermocellum* are a few of them. Many facultative and obligate anaerobes produce lactate as a common byproduct of fermentation. Lactic acid bacteria (LAB) are bacteria that generate lactate as a primary fermentation byproduct. The majority of LAB are only partially able to synthesize the monomers required for biosynthesis and the vitamins that are growth factors. Although LAB are thought of as obligate anaerobes, they can consume oxygen and, in the presence of hemin in the medium, can synthesize cytochromes. While some LAB simply generate lactate from carbohydrates, others additionally generate ethanol and acetate. Sugars are fermented by homofermentative LAB via the EMP pathway and heterofermentative LAB via the phosphoketolase pathway. The majority of Lactobacillus species, as well as Sporolactobacillus, Pediococcus, Enterococcus, and Lactococcus, are homofermentative LAB. They produce ATP by using hexoses and the EMP pathway. Using pyruvate as the electron acceptor, lactate dehydrogenase reoxidizes the NADH that was reduced during the EMP process(Taillefer & Sparling, 2016). The pH inside cells decreases as lactate builds up during fermentation. In acidic environments, lactate dehydrogenase is active and produces lactate as the main product. Large amounts of ethanol and acetate are produced by homofermentative LAB in an alkaline environment. In addition to lactate, some species of Bifidobacterium and Leuconostoc create ethanol and acetate. They use the phosphoketolase pathway, a distinct glycolytic mechanism. Heterofermentative LAB, such as *Leuconostoc mesenteroides*, oxidize glucose-6-phosphate to ribulose-5-phosphate. Before being cleaved by phosphoketolase into glyceraldehyde-3-

phosphate and acetyl-phosphate, epimerase converts ribulose-5-phosphate to xylulose-5-phosphate. Similar to homolactate fermentation, glyceraldehyde-3-phosphate metabolism results in lactate, which produces ATP. In the process of oxidizing glucose-6-phosphate, acetyl-phosphate is reduced to ethanol, which serves as an electron acceptor to oxidize the NADH that has been reduced. This fermentation produces one ATP per hexose.

Without reducing NAD^+, pentoses are transformed into xylulose-5-phosphate. Here, acetyl-phosphate is employed in the synthesis of ATP rather than as an electron acceptor. *Leuconostoc mesenteroides* produce two ATP from a pentose molecule and one ATP from a hexose molecule. Using two distinct phosphoketolases, one active on fructose-6-phosphate and the other on xylulose-5-phosphate, Bifidobacterium bifidum ferments two molecules of hexose to two molecules of lactate and three molecules of acetate. This bacterium uses two glucose molecules to create five ATP. Acetyl-phosphate is utilized to create ATP in pentose metabolism by *Leuconostoc mesenteroides* because hexose-6-phosphate is not processed by a reductive mechanism. Because they are unable to synthesize them, LAB needs a variety of vitamins, nucleic acid bases, amino acids, and other chemicals as growth factors. They are able to create a small number of monomers and polymers from acetyl-CoA. In the fermentation of homolactate, neither acetyl-CoA nor acetyl-phosphate operate as intermediaries. The process by which pyruvate is converted to acetyl-CoA varies according on the strain. Similar to aerobic bacteria, most Lactococcus and Enterococcus species oxidize pyruvate to acetyl-CoA by the action of the multienzyme complex Pyruvate Dehydrogenase. Aerotolerant obligate anaerobes, LAB are equipped with superoxide dismutase and peroxidase for the detoxification of superoxide (O^{2-}) and hydrogen peroxide (H_2O_2). From sulfite, peroxidase produces sulfite radicals, which are lethal to cells. Therefore, LAB from peroxidase-negative aerobic microbes can be selectively removed using sulfite. When hemin is supplied, the catalase activity of several LAB, including *Enterococcus faecalis*, *Lactobacillus brevis*, and *Lactobacillus plantarum*, changes from negative to positive. These species lack hemin, but they do possess the genetic information necessary to synthesize the catalase apoprotein.

7.11PEROXIDASE

Peroxidase, also known as oxidase, boosts cell production in addition to ETP by oxidizing NADH and using O_2 as an electron acceptor(Jurtshuk et al., 1975). In heterofermentative LAB, when peroxidase oxidizes NADH, acetyl-phosphate that is not required as an electron acceptor can be used as a substrate for substrate-level phosphorylation to produce additional ATP. Acetate kinase is the catalyst for this process. Similar to this, when H_2O-forming NADH oxidase oxidizes NADH to NAD^+ reducing O_2, homofermentative *Lactococcus lactis* oxidizes pyruvate to acetate or condenses it to acetoin instead of utilizing it as an electron acceptor. *Lactococcus cremoris* produces a proton motive force when its internal lactate content exceeds that of the medium. The process known as malolactic fermentation is the conversion of malate to lactate by *Lactobacillus plantarum*, *Lactobacillus casei*, *Leuconostoc mesenteroides*, *Leuconostoc oenos*, and *Lactococcus lactis*. In the presence of fermentable substrates like glucose, malolactic fermentation occurs because these bacteria are unable to use malate as a source of carbon and energy. During the carboxylation of malate to lactate and the exchange of malate/lactate, a proton motive force is produced in this fermentation process. When glycerol is dehydrated, 3-hydroxypropionaldehyde is produced, which is used by heterofermentative Lactobacillus species as an electron acceptor. Because these bacteria preferentially use this electron acceptor, pyruvate is oxidized to acetate through an acetate kinase-catalyzed process that produces ATP. As a result, heterofermentative LAB grows more readily on glucose plus glycerol than it does on glucose alone.

Chapter 8

ECOLOGY AND MICROBIAL PHYSIOLOGY

Contents

8.1 INTRODUCTION

The current consensus is that life began as self-replicating molecules in the aquatic environment between 3.5 and 4 billion years ago(Das, 2022). Microorganisms that are highly adapted to aquatic conditions came into being as a result of the evolution of prokaryotes and eukaryotes. There are many possible benefits that aquatic life has over terrestrial life. These include minimal water loss, passive movement by water currents, physical support, lower temperature, and readily available soluble organic and inorganic nutrients. Osmotic variations between the organism and the surrounding aquatic medium, which can result in endosmosis or exosmosis, and a high degree of physical disturbance in many aquatic systems are potential disadvantages of aquatic habitats. For light to be available, photosynthetic organisms in undisturbed aquatic systems like lakes must continue to occupy the top of the water column. Physical and chemical characteristics of many water bodies such as the water column of lakes exhibit a variety with few discrete microhabitats. In these conditions, species compete based more on divergent development and reproduction strategies than on particular environmental adaptations to a certain location. Aquatic sites exhibit a gradient from freshwater, which has a low ionic content, to saline, which has a high ionic content and is usually dominated by sodium and chlorine ions. Estuaries, saline/salt lakes, and large areas of the polar ice caps are examples of saline waters. High osmotic potential and high electrical conductivity are indicators of high ionic concentrations.

8.2 PHYSIOLOGICAL REQUIREMENTS

Because freshwater and saline environments have diverse physiological requirements, aquatic organisms often live in either freshwater or saline environments because they are adapted to one environment but not the other. A recent assessment of saline lakes in Australia revealed the significance of salinity in defining the species composition of the aquatic microbial community. The diatoms in the high salinity (hypersaline) and low salinity (oligosaline) waters were found to be different. However, other diatom species were found in all saline conditions, suggesting that certain microorganisms are fully unaffected by ionic ratios and salt content. The capacity of certain species to migrate from saltwater to freshwater areas and establish themselves in their new locations is another sign of long-term adaptation to various saline conditions. This seems to be the case for several freshwater littoral red algae, which originated in marine habitats. Disparities in microbial communities between freshwater and saltwater habitats are especially important on a global scale, where saline conditions predominate in terms of total biomass, and overall contribution to carbon cycling. The physical/chemical properties of freshwater environments vary greatly, as does the impact these features have on the microbial populations inhabiting there. These factors are taken into consideration, but first, it is critical to distinguish between the lentic and lotic systems. The distinction between lotic and lentic systems is not perfect. Benthic communities range in type from net heterotrophy found in sediments where anaerobic heterotrophic bacteria predominate to net autotrophy found in shallow lake sediments where benthic algae predominate. Pelagic communities often contribute more to autotrophic processes in lentic water bodies and receive more light exposure than benthic systems. It is vital to maintain the balance of available nutrients; low-nutrient (oligotrophic) lakes usually exhibit net heterotrophic behavior, whereas high-nutrient (eutrophic) lakes exhibit net autotrophic behavior.

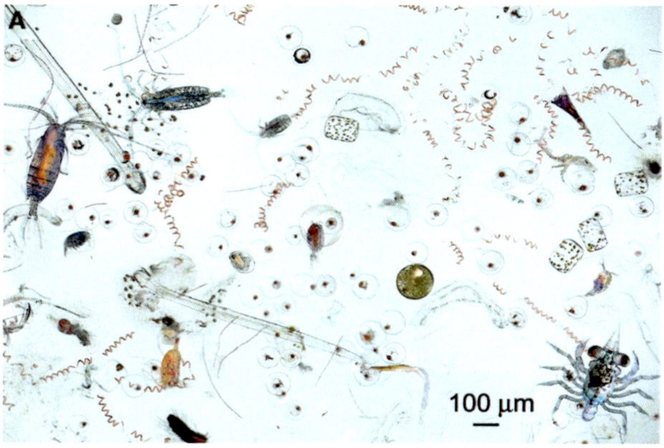

Figure 8.1 Plankton

Source: https://upload.wikimedia.org/wikipedia/commons/2/2f/Marine_micro-plankton.jpg,*Creative Commons License*

8.3 PLANKTON

Allochthonous carbon intake can still lead to substantial heterotrophy(Rautio et al., 2011). Both forms of nutrition are more prevalent in eutrophic lakes, but autotrophy is strongly encouraged by the increased availability of soluble inorganic minerals. Lotic pelagic communities may likewise exhibit wide variations in net ecosystem productivity. This is especially evident in certain estuaries, where there is an obvious seasonal variation in the net ecosystem productivity or NEP value about river inflow. Based on where they spend the majority of their growing phase, freshwater organisms can be classified into two broad groups: benthic organisms, which are found in sediments, and pelagic species, which are found in the main body of water. Plankton, which is free-floating, and nekton, which are strongly swimming animals like fish, are some of the pelagic biotas. Although they may have restricted motility and be able to move vertically, the latter tend to swim within the body of water. Since plankton (Figure 8.1) makes up the majority of pelagic microorganisms, the difference in this group is between planktonic and benthic states. While planktonic and benthic phases are distinct, most species exhibit both planktonic and benthic phases throughout their life cycles. For instance, planktonic and attached states are dynamically balanced in biofilm bacteria, while planktonic/benthic interactions are evident in freshwater algae found in lake and stream habitats.

The metabolic activity of the same species in planktonic and benthic phases usually varies significantly. Planktonic organisms are further classified into meroplanktonic species where the planktonic phase is constrained and holoplanktonic forms where the organism is present in the water column for a substantial part of the life cycle. The algae are a particularly good example of the difference between holoplanktonic and meroplanktonic species, with the meroplanktonic condition being thought of as an adaptation. Biota, which includes bacteria, fungi, and protozoa that can decompose organic detritus found in sediments, are the dominant organisms found in these environments, both at the surface and within sediments.

Certain benthic species have been defined by a variety of key concepts, one of which is the periphyton. The type of substrate, which might be alive (plant surfaces) or nonliving (organic or inorganic, with varying particle sizes), causes differences within the periphyton. Many bacteria remain dormant over the winter on the sediments, especially in temperate areas. When environmental factors worsen such as decreased availability of nutrients and light, buildup of hazardous metabolites, and decreased temperature, resistant spores usually develop first. Sexual reproduction may also occur before this happens. Dormancy sometimes seems to be connected to a specific shift in the environment, like the amount of oxygen present.

The photosynthetic purple bacteria are metabolically active in the anaerobic hypolimnion of the lake water column during stratification. Metabolic inactivity in bacteria and viruses is also connected to nutrients. A comparison of viable and total bacterial counts indicates that the proportion of metabolically active aquatic populations is small. Bacteria in certain environments stay dormant until they come into suitable growth conditions, especially about the availability of nutrients. Moreover, viruses do not metabolically function in the aqueous medium; they become active only in the presence of viable host cells. They exist in the aqueous medium as free particles or virions.

8.4 AQUATIC ENVIRONMENTS

Major aquatic environments include wetlands, rivers, and lakes. Together, they constitute a distinct biological unit with a unique community of organisms. The pelagic zone of neuston, phycosphere, and photosynthetic bacterial communities can be further divided(Camacho-Chab et al., 2016). They exist in distinct, small-scale settings and are made up totally of microbes. The associated organisms that make up the littoral zone are primarily classified into three broad groups in eutrophic lakes: the communities of macrophytes, upper periphyton, and lower periphyton. These contain a variety of unique micro-communities, such as bacterial and algal biofilms on exposed rock surfaces and epiphytic communities on the leaves of macrophytes. These different groups of organisms are symbiotic within the larger ecosystem. For example, during the summer, the benthic community may appear to be very different from the pelagic community. However, the benthic community depends on a constant supply of biomass from the pelagic zone through sedimentation, and populations of planktonic organisms are formed through recruitment from the benthic zone. Every community has a unique pattern of living things with varying degrees of biodiversity. This diversity within these communities can be quantified in many ways, such as the proportions of autotrophic and heterotrophic species, the presence of mucilaginous and non-mucilaginous types, the size range of organisms, and the attached versus free-floating biota. The species content is frequently used to define biodiversity. Algal diversity within phytoplankton is taken into account when calculating species diversity indices. These indicators show significant seasonal variations in phytoplankton development, suggesting shifts in community diversity throughout the growth season.

Analytical microscopical techniques have enough spatial resolution to identify the chemical and molecular composition of individual microbes within mixed populations. Molecular methods have been used to investigate many facets of genetic variation within natural populations and offer great potential for examining biodiversity within populations of a single species. These include alterations in plasmid-borne resistance in aquatic bacteria and strain variance in blue-green algae. Within benthic environments, variations in the expression of stationary phase genes between adherent (biofilm) and planktonic bacteria represent an important intraspecific variable. Ecosystems are home to a diverse spectrum of living things that engage in interactions with the environment. A single ecosystem has several significant characteristics, including preservation of the internal environment, well-defined pathways for biomass formation and transfer, interactions with the external environment, and interactions between species. The variety of species found in aquatic ecosystems contributes to the creation and exchange of biomass as well as defining and characterizing the system. They occupy certain trophic levels and constitute an interconnected system of relationships known as the food web. They have distinct interactions within the ecosystem. The primary factors influencing the equilibrium of individual species in the food web are competition and resources, such as light and nutrient availability. This has significant effects on biodiversity which in turn impacts how well the system functions as a whole. Physical stress levels in the environment and the stability of the ecosystem are intimately linked to community organization. Ecosystems interact in a variety of ways. The net carbon exchange between the surrounding atmosphere and the aquatic ecosystem is quantifiable in terms of net ecosystem productivity (NEP).

8.5 PELAGIC ECOSYSTEM

The pelagic ecosystem is a large-scale functional unit that contains a diverse spectrum of biota, whereas the microbial biofilm is a small-scale ecosystem made up only of microorganisms. In the case of biofilms, comparatively less is known regarding food webs.

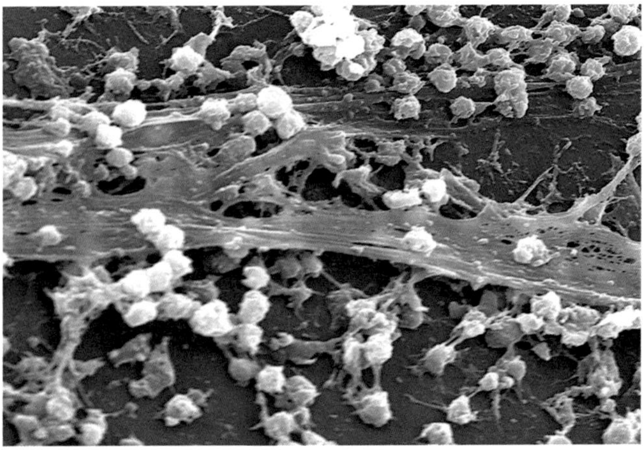

Figure 8.2 Microbial biofilms

Source: https://upload.wikimedia.org/wikipedia/commons/2/2f/Marine_micro-plankton.jpg,*Creative Commons License*

Microbial biofilms (Figure 8.2) offer a system for us to think about the essential elements of ecosystem function and community interactions(Van Colen et al., 2014). Due to the intimacy of the organisms within the biofilm, there is a high level of biological interaction, and their small scale makes them suitable for laboratory as well as environmental experimentation. Microbial biofilms are found as a surface layer on rocks in lakes and rivers. They are distinct communities inside a gelatinous matrix. Environmental elements including water flow rate, ambient light intensity, and past colonization history all affect the biological composition of biofilms. These can be completely bacterial in some circumstances, or they can start as diatoms that settle and grow into bigger periphyton groups.

The biofilm is an established mixed biofilm, the kind that could grow in low-light conditions on a river substratum. It is made up of a well-balanced colony primarily of bacteria, with some protozoa, fungi, and algae as well. The majority of the species are found in a gelatinous matrix that determines the ecosystem size. With water seeping through the channels or pores between the columns, the matrix usually looks like a column.

The microbial biofilm shares the four main components of ecosystem function. They are interactions between organisms, biomass transfer, homeostasis, and interactions with the external environment with larger ecosystems as well as other small-scale microbial ecosystems, such as phycosphere and neuston. Microbial biofilms are not isolated from their biological or physicochemical environments, despite functioning as a functional and self-contained entity. Light, which is necessary for photosynthesis, inorganic nutrients, soluble organic nutrients, and dissolved oxygen are important physicochemical properties. Internal water circulation facilitates the entry and exit of soluble components from the biofilm and creates concentration gradients within the gelatinous matrix.

These gradients in concentration might have an important role in determining the physiology of local biofilms. Laboratory studies on monospecies biofilms show that the concentration and quality of nutrients have a major influence on biofilm structure, indicating the potential importance of nutrients. Biofilms are directly impacted by the external biotic environment. These are mediated by the grazing activities of invertebrates as well as by the influx and egress of resident microorganisms at the water interface.

Another biotic process that contributes significantly to the substrates needed for heterotrophic nutrition is the introduction of soluble or particulate organic materials into the biofilm. Dissolved organic carbon, or DOC, is essential for connecting bacteria to phytoplankton and zooplankton productivity. Exogenous DOC also serves as a bridge connecting the pelagic ecosystem to the outside environment. Energy (light), carbon dioxide, and inorganic nutrients enter into phytoplankton, which can be further classified into a range of subgroups. Protozoa and various zooplankton groups subsequently ingest these in diverse ways, which are subsequently consumed by zooplanktivorous fish, which in turn are consumed by piscivorous fish and other vertebrates as apex predators. Carbon flow represents trophic interactions between groups.

8.6 HOMEOSTASIS

All ecosystems must maintain homeostasis or the internal environment, and this is especially true of the pelagic environment(Egorov, 2021). Homeostasis refers to the system's capacity to withstand transient physical, chemical, and biological perturbations. These perturbations might originate internally (random growth in a microbial population) or be imposed from the outside (severe climatic change and pollutants). Homeostasis functions to preserve the ecosystem's biodiversity and to bring the biotic ecological balance back to normal. Apart from its significance for organic processes in nature, homeostasis plays an essential role in the way a community reacts to the nutrient enrichment (eutrophication) of the freshwater ecosystem caused by human activity.

The breakdown of ecosystem homeostasis leads to the detrimental environmental effects of eutrophication, and the recovery of freshwater systems necessitates the return of homeostatic mechanisms and the restoration of a balanced community. Apart from the impacts of seasonality, diurnal periodicity in phytoplankton activities is also encouraged by external variations in light intensity. These include the entrainment of physiological processes (photosynthesis, nitrogen fixation), dynamic activities (vertical migration) of phytoplankton, and gene transcription (expression of clock genes). Daily variations in light intensity also have a direct impact on other biota, such as zooplankton, and diurnal periodicity encourages these vertical migrations and interactions with phytoplankton. These include the time and location of phytoplankton grazing by zooplankton (inside the water column) and the phosphorus recycling process between phytoplankton and zooplankton. With the input and outflow of soluble nutrients, particulate matter, and aquatic biota, the dynamics of lake hydrology may also have a significant external impact on the pelagic ecology.

For example, inflow may be especially crucial toward the end of the growth season in providing soluble inorganic nutrients to a nutrient-depleted system, whereas outflow of biota, such as planktonic algae, may have an impact on bloom development. When it comes to many freshwater systems, flushing the pelagic system is a crucial management tool for controlling blue-green algal blooms that appear in late summer. The lake surface may also be a site of biota recruitment and loss. While insects like mosquitoes and chironomid flies enter the system, go through an aquatic larval stage, and then leave the environment as they metamorphose into the imago stage, this does not seem to have been reported for lake microorganisms. Environmental changes negatively impact the functionality of all biological systems. These are known as stress factors and can affect molecular systems, individual species, and ecosystems.

8.7 PROKARYOTES

Prokaryotes induce certain transcription factors, known as sigma proteins, in response to physicochemical stresses like salinity, high temperatures, and sudden food shortages (Seshasayee et al., 2011). These factors cause the expression of a variety of stress genes. The biological community frequently changes as a result of internal or external environmental effects on ecosystems. When these effects affect the dynamic aspects of the biological system, such as food webs, they might be classified as environmental stress. One crucial factor in lotic systems that influences the growth of biofilm and epilithon communities is the water flow rate. Another illustration of external stress is the effluent flow rate into activated sludge (sewage) systems. The protozoon community at high flow rates is restricted to species that reproduce quickly. The introduction of inorganic fertilizers into lakes and other standing waters is another important stress that contributes to eutrophication and the disruption of ecological equilibrium. Internal stress factors that emerge in aquatic situations could potentially be important. These internal stresses can occasionally be subsequent outcomes of external causes, such as eutrophication-induced Bluegreen algal blooms.

The production of anoxic conditions by the bacterial bloom that arises in reaction to organic pollution is another example of internal stress that may be caused by external sources. This affects the developing protozoon activity and composition significantly. In other situations, there is no clear connection between internal stresses and outside disturbances. They can be a part of the seasonal cycle in lakes, and they include the rise in zooplankton grazing activity that takes place in the clear-water phase, which affects the phytoplankton population.

One example of homeostasis or maintenance of the internal environment, is the capacity of ecosystems to withstand stress and return to, or retain, their biological integrity. The biochemical, and physical properties of an ecosystem make up its internal environment. An ecosystem's stability is determined by how much it is impacted by stressors; the more stable the system, the more resilient it is to outside/external change. Many of the concepts related to ecosystem stability and the mechanisms controlling it can be applied to aquatic systems, even though they were primarily established in terrestrial ecosystems. The ecosystem stability is primarily related to biotic interactions and the food web applies to established terrestrial systems, such as tropical rain forests, even if certain freshwater systems have evolved highly stable and complex populations. These generally follow the community structure concept, which gives less importance to physical disturbance. Competition is mostly between predators and primary producers. Predation keeps herbivore populations below the carrying capacity (theoretical) of their available food source, preventing them from restricting the growth of primary producers. Freshwater ecosystems are often physically heterogeneous and subject to more physical disturbance than their terrestrial counterparts. These systems resemble a community structure model, which postulates that significant changes in physical stress (disturbance) lead to changes in competition and predation. Maximum stress prevents the emergence of any species, but as stress decreases, the community grows more complex until it reaches low-stress levels. Both lotic and lentic aquatic environments exhibit differences in stress levels and community structures. Rapidly flowing rivers with turbulent water offer a suitable illustration of a high-stress environment, which corresponds to the Menge-Sutherland model. Unattached species, like invertebrates and microbes, are unable to establish themselves in this environment and are carried away by the current. These conditions are only sustained by attached microorganisms, such as biofilms and periphyton communities, with different levels of competition among species populating exposed surfaces.

8.8 ALGAE

A diverse group of aquatic organisms known as algae are capable of photosynthesis. Algal succession may not develop in severely disturbed environments, like headwater streams(Blum, 1957). Stress factors have an impact on the microbes of lakes in both natural and artificial environments. In natural conditions, seasonal cycles are characterized by changes in stress levels and community structure. During the clear-water phase, when phytoplankton populations are low and subject to quick fluctuations, there is intense competition among zooplankton, the herbivores, and high stress levels are experienced. Now, the only organisms that can withstand the harsh environment are algae which develop quickly. On the other hand, the summer algal bloom period is usually a low-stress period during which significant levels of competition arise between established populations of different phytoplankton species and zooplankton grazing pressure is decreased. Diverse stresses have diverse effects on communities in terms of their intensities and types. These reactions are also known as biological response signatures. They can be utilized to assess the cause, nature, and effects of stress on a specific aquatic system. Observable patterns in the response of aquatic communities that enable the investigator/researcher to distinguish between various stressors are known as biological response signatures. The ecology of standing water (lentic) systems is dominated by phytoplankton, and photosynthesis is the primary internal source of carbon entering the food chain. On the other hand, carbon enters moving streams mostly by external sources (allochthonous origin) as complex organic molecules, with attached (benthic) organisms predominating(Lock, 1994).

Because of these distinctions, lotic systems usually exhibit net heterotrophy, whereas lentic systems frequently exhibit net autotrophic metabolism. In terms of microbiology, biofilm communities are especially significant in rivers and streams. Depending on their location and developmental stage, these communities are dominated by either bacteria or algae. Throughout the life cycle, there are significant variations in the concentrations of dissolved and particulate carbon in many streams and rivers. Recorded dissolved organic carbon (DOC) values typically fall between 1 and 10 mg L^{-1}, but in certain rivers, they can reach significantly higher levels. The seasonal variations in entry into the river system and the allochthonous origin of these chemicals are reflected in most cases. For instance, the DOC content in Southern Asian rivers varies greatly, and seasonal flow patterns and the timing of DOC maxima and minima are closely correlated. The Indus and Ganges-Brahmaputra Rivers get maximum dissolved organic carbon levels towards the end of rising water levels, which is caused by entrainment and overflow from highly productive floodplains. Then, when water levels drop, the pulse of DOC rapidly decreases as a result of mixing and dilution. The extremely high autumn allochthonous DOC levels in the upper Mississippi River (USA) are caused by leaf litter leaching rather than soil erosion. Seasonal trends can range greatly, even though many rivers exhibit increased allochthonous DOC concentrations during floods and runoff. For instance, the Shetucket River in the United States has very low DOC levels during the winter months when inflow is highest and maximum levels during the summer months when inflow is lowest. The production of autochthonous carbon by the senescence and secretion of benthos was suggested as the cause of the summer maxima. Because of ice melt and strong rains in the winter, the allochthonous DOC intake was diluted by high discharge.

8.9 BIOLOGY OF ALGAE

Some river seasonal rhythms are influenced, at least somewhat, by planktonic primary production occurring in situ. While autochthonous DOC accumulates during the low-flow phase, allochthonous DOC concentrations in the Gambia River (West Africa) increase during the moment of maximum discharge into the system. This was connected to the production of phytoplankton and to higher pH values in river water. Phytoplankton in most lotic systems are only carried along by the current; they are unable to establish stable communities. Net production within a specified segment is only possible when local growth rates surpass downstream losses due to this constant displacement. In lotic conditions, turbidity, circulation within the water column (no stratification), and ambient light intensity (hanging vegetation) tend to be the main factors limiting phytoplankton development, whereas current velocity plays a major role in downstream loss. The formation of a pelagic community and the production of phytoplankton in riverine systems are thus primarily controlled by light availability in conjunction with hydrological processes. Phytoplankton production in large rivers may be significant. When phytoplankton is exposed to favorable light conditions for an extended period of time, at peaks of low discharge and turbidity, local biomass buildup occurs mostly in shallow reaches. The pool phytoplankton productivity was less than 10% of the phytoplankton input from tributary and upstream sources. Phytoplankton production in the pool surpassed that of external algal sources during low discharge periods. These take place on specialized photosynthetic (thylakoid) membranes and entail the conversion of light energy into adenosine triphosphate (ATP) and NADPH by photosynthetic pigments. Water splits during the process of oxygenic photosynthesis in eukaryotic and prokaryotic algae, releasing oxygen and providing electrons needed to change $NADP^+$ into NADPH. Since light reactions are photochemical in nature and not enzymatic, they are temperature-independent. There are two primary types of light reactions: photosystem II (PS II), which splits water, and photosystem I (PS I), which is linked to the conversion of NADP to NADPH(Lubitz et al., 2019). The light-harvesting complex, a macromolecular structure found in every system, is split into two distinct regions: an internal reaction center and a surface antennal complex. The antennal complex surface pigments are responsible for absorbing light energy, which is then directed into the reaction center. To facilitate the transport of electrons to electron chains, this molecule of chlorophyll-a (P_{680}) is present as a single molecule. Chlorophyll-a is one of the many accessory pigments found in antennas, along with phycocyanins, phycoerythrins, carotenoids, and other pigments.

Both photosystems are necessary for photosynthesis, but they differ in terms of their physical composition, photon requirements, and antennal pigment types. For instance, chlorophyll a is primarily linked to PS I in red and blue-green algae, but phycoerythrins are primarily found in PS II harvesting complexes. These processes, which take place in the gaps between thylakoid membranes, use the energy and reducing the power of ATP and NADPH to convert CO_2 to hexose, which yields a range of sugars, amino acids, and fatty acids. Algae are widely distributed within specific ecosystems, existing as both benthic and pelagic forms, while photosynthetic bacteria are usually confined to small microhabitats. Algae are physiologically more diverse than prokaryotes because they have a variety of anatomical and physiological characteristics. For photosynthetic organisms, light is essential for both general growth (ATP and glucose synthesis) and more specialized processes like gamete differentiation. For instance, in order for green algae to reproduce sexually, light is essential.

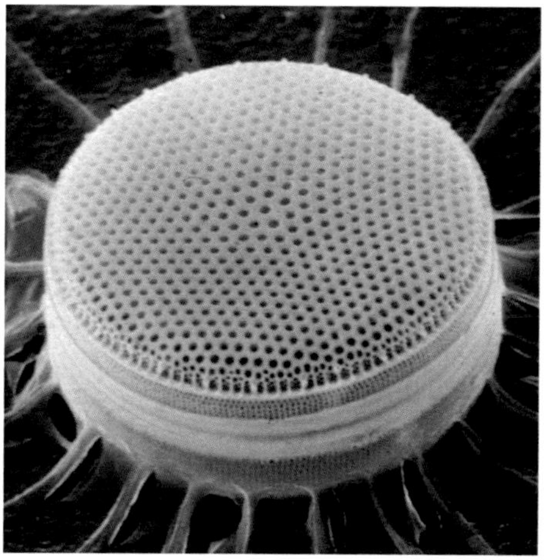

Figure 8.3 Diatoms

Source: https://upload.wikimedia.org/wikipedia/commons/b/b3/Detail%2C_CSIRO_ScienceImage_7632_SEM_diatom_%28cropped%29.jpg,*Creative Commons License*

8.10 ENERGY MECHANISMS OF ALGAE

Algae exhibit adaptations related to energy conservation or efficient utilization of the energy converted during photosynthesis. The synthesis of macromolecules for growth uses up a large portion of this energy. When many algae are compared, it becomes clear that there are wide variations in the energy needed to build cell walls. Because the silica cell wall (diatom) (Figure 8.3) costs roughly 12 times less energy to build than the cellulose and peptidoglycan cell walls of other algae, these organisms can surpass other algae in terms of energy. This could explain these organisms' evolutionary success as well as the fact that diatoms are the most prevalent type of algae in both fresh and saltwater environments. However, there is a trade-off between the high mass and silicon demand associated with diatom energy savings. Due to the requirement of silicon, the expansion of diatom populations may abruptly stop when silicon supplies are lowered. Due to the high specific gravity and sedimentation rate caused by the high mass of diatom cell walls, many species are confined to specific types of lakes and require turbulent water conditions to remain in suspension. The total lack of a cell wall reduces building or construction energy. This method uses minimal energy in freshwater environments because the naked cell must waste a significant amount of energy to maintain its osmotic balance through the use of a contractile vacuole. Using energy as efficiently as possible is crucial for movement within the water column. Molecules that swim actively, like dinoflagellates and cryptophytes, must constantly expend energy(Sommer, 2023). Energy is needed to create the gas vacuole proteins in organisms like blue-green algae, which employ buoyancy to control their location in the water column. Once these proteins are produced, regulation occurs through the deposition and removal of carbohydrates. Less additional energy is needed for the diurnal migration in the water column because this is a typical aspect of these. Depth regulation in response to transient variations in light intensity will likewise use less energy because the creation of ballast at the lake surface in strong light conditions will also sink the cells out of the zone of photoinhibition. Diatoms depend on water turbulence rather than direct energy expenditure to maintain their location in the water column because they lack buoyancy and an active motility mechanism.

In temperate lakes, competition with other algae and the physical/chemical environment determine how successful each of these methods is during the annual cycle. During the relatively static conditions of the stratified water column in the summer, organisms with buoyancy regulation or active motility tend to replace diatoms, which are adapted to turbulent water conditions from autumn to spring. Dinoflagellates are an example of an actively motile organism. While they need a lot of energy to travel through the water column, their ability to move at night from a nutrient-depleted epilimnion to lower parts of the hypolimnion for nutrient uptake gives them a competitive advantage in late summer. Although blue-green algae can migrate into the hypolimnion, active photosynthesis is necessary for the creation of ballast in their buoyancy mechanism. Due to lower light levels in the autumn, this may become restricted, which would subsequently allow dinoflagellates to take over. The production of low molecular weight solutes involved in osmoregulation is intimately related to photosynthesis.

Particularly when simple sugars are formed, a variety of osmotically active chemicals are produced that can be used to identify specific algae species. These substances play a critical role in maintaining the osmotic equilibrium in freshwater, where a hypotonic solution envelops the algal cells. Under these conditions, the tendency of water to enter through endosmosis is either balanced by the pressure of the cell walls or, in the case of naked algae, necessitates the constant ejection of water by contractile vacuole activity. Higher intracellular quantities of osmotically-active solutes are needed to balance the increased external molarity in conditions of rising salinity, where the external medium becomes hypertonic. The diatom *Cyclindrotheca fusiformis* serves as an example of the significance of light in the osmoregulatory process(Van Bergeijk et al., 2003). Photosynthesis keeps the free internal mannose concentration at a sufficient level to balance external osmolarity. It is possible to produce free mannose from its polymer, polymannose, in the dark. Under light conditions, free mannose is converted to polymannose when the external molarity decreases. Raising the external molarity prevents the synthesis of polymannose and causes the direct synthesis of free mannose by CO2 fixation.

8.11 ASSESSING LEVELS

There are some methods for measuring epipelon biomass. In the majority of investigations on freshwater systems, an algal sample is obtained by covering the exposed sediment surface with lens paper. By using this method, estimates of epiphytic algae typically result in very low biomass levels. These low values partially reflect the shading that wetland epipelon experience from floating metaphyton mats, phytoplankton blooms, submerged macrophytes, and their associated epiphytes, all of which lower light levels at the sediment surface during the growth season. Core sample chlorophyll-a analysis yields results that are significantly higher and more akin to salt marsh epipelon. Salt marshes may yield higher epipelon biomass concentrations. Because plant surface area is correlated with epiphyton biomass, it is typically not generally stated per unit area of wetland. The majority of data are often based on algae that are adhered to morphologically simple artificial substrata; in freshwater marshes, biomass levels are typically less than 5 μg cm-2 of chlorophyll-a. According to studies that have calculated epiphyton biomass per unit wetland area, this biomass is comparable to the amount of algae seen in other shallow waters. Since dry weight is typically used to convey the minimal amount of quantitative data on metaphyton, comparisons with other algae are challenging. The biomass of metaphyton is typically less than 10 mg cm-2, however, it can reach significantly high values when it is growing quickly and generating large surface mats. Blue-green algae can detect and react to variations in the amounts of nutrients and oxygen in addition to light. It appears that cyclic nucleotides, such as cyclic guanosine monophosphate or cGMP and cyclic adenosine monophosphate (cAMP), are important for signal transduction. Anabaena and other blue-green algae exhibit quick environmental reactivity, with cAMP concentrations increasing during light-to-dark transitions and vice versa(Borbély et al., 1990). These findings imply that in blue-green algae, cAMP functions as a secondary messenger of light signal transduction, regulating the activity of many processes that are sensitive to light.

8.12 BLUE-GREEN ALGAE

A class of tiny, photosynthetic bacteria known as cyanobacteria, or blue-green algae, is present in many different types of water bodies.

Blue-green algae (BGA) typically have similar amounts of the secondary messengers, cAMP and cGMP, under normal growth conditions, but variations in light, nutrients, and oxygen cause variations in the amounts of both nucleotides. Cyclic nucleotide phosphodiesterases control intracellular concentrations and are essential for the transmission of cAMP and cGMP signals. Using bioinformatics and protein sequence-function studies, whole genome sequences of the unicellular Bluegreen algae revealed phosphodiesterases. A group of regulatory genes surrounds these phosphodiesterase genes, and all of the genes are transcribed together. The polypeptides encoded by the regulatory genes function as two-component systems, with the receiver (response regulator) and transmitter (sensory histidine kinase) domains. Signaling motifs found in these genes are known to control light-stimulated signaling pathways in other animals, especially those that include blue light. Blue light is known to alter the intracellular concentration of cAMP in Synechocystis, indicating that the putative cAMP phosphodiesterase is involved in the transduction pathway of the blue light signal. Through the modulation of gene expression, signal transduction of variations in light intensity enables cells to adapt to environmental changes and produce an integrated response. Many metabolic processes in freshwater algae are impacted by light-modulated gene expression. Although the light-responsive production of photosystem proteins and pigments is particularly pertinent to photoadaptation. Because of their light-modulated activity, algae can adapt to changes in light levels in the changing aquatic environment, maximizing their photosynthetic output over a short amount of time. By encouraging the production of proteins that bind to chlorophyll-a, photosystem II center proteins, and phycobilin pigments (phycocyanin and phycoerythrin), this is achieved. Different genes exhibit varying degrees of wavelength selectivity for the expression of light-responsive genes, indicating the involvement of different receptors and multiple transduction pathways. For certain genes, the function of the final product is directly correlated with the activation wavelength. For instance, in the blue-green alga Calothrix, red light stimulates phycocyanin synthesis, but green light activates phycoerythrin synthesis.

8.13 GENE EXPRESSION

Light-responsive gene expression and protein induction are not fast enough to respond to short environmental changes in some light-dependent physiological processes. For instance, in nitrate absorption, a more sensitive system of reversible enzyme inhibition/activation has also emerged. Lake macrophytes are mostly confined to the littoral zone, or the margin of the lake, or other shallow areas in temperate climates, where they face minimal competition from phytoplankton for light. As the littoral zone moves from an oligotrophic to a eutrophic state, macrophytes grow abundantly; nevertheless, excessive nutrient concentrations cause strong algal blooms and suppress higher plants. The formation of floating macrophytes, such as water lettuce (*Pistia stratoites*) and water hyacinth (*Eichornia crassipes*), is common in tropical and subtropical lakes(Lu et al., 2010). Shallow pools are examples of wetland settings where macrophytes have developed extensively. These plants can outcompete algae and dominate primary production. Both anchored and floating macrophytes have the potential to intercept a substantial amount of light. The growing canopy can exclude up to 95% of the incident radiation. Light transmission through floating Lemna mats might only account for 0.1% of surface irradiance. Seasonal wetlands exhibit macrophyte dominance. For instance, the rainy season in North West Australia brings high humidity, thunderstorms, and torrential rain, resulting in extensive flooding over large areas of land. A diverse array of quickly expandingg hydrophytes is supported by these wetlands, but they eventually go dormant as the dry season approaches.

Figure 8.4 *Ceratophyllum demersum*

Source: https://upload.wikimedia.org/wikipedia/commons/0/05/Ceratophyllum_demersum.jpg,*Creative Commons License.*

Algae have a competitive advantage when pH levels rise because they are generally more effective at assimilating bicarbonate ions than higher plants. Furthermore, not all higher plants exhibit this action. *Ceratophyllum demersum*,

(Figure 8.4) a hydrophyte, has two varieties: one adapted to light, and the other to shade, with varying levels of bicarbonate uptake in both. The light-adapted cultivars can withstand pH variations through bicarbonate and free CO_2 absorption, and they can coexist in competition with algae in high-light conditions. Shade-adapted cultivars can only assimilate free CO_2 and can only grow in waters with a relatively low pH. Higher plant uptake of CO_2 causes pH levels to rise to 11.5. Only bicarbonate consumers, such as *Elodea canadensis*, all species of Potamogeton can survive at this level, which is characterized by the presence of bicarbonate and carbonate ions but the absence of free CO_2. For many higher plant species in eutrophic and hypereutrophic situations, the absence of free CO_2 is a problem, and macrophytes that depend on free CO_2 can no longer undergo photosynthesis at pH 8.5–9. One crucial component of algal-macrophyte competition is the rapid development early in the season. Vascular photosynthetic organs accumulate at or near the water surface which gives them the ability to compete with or even inhibit phytoplankton. The evolutionary development of surface-floating leaves or partial shoot emergence/development gives them a competitive advantage over algae in the struggle for light and CO_2. The steep oxygen gradient in the water column, with high levels at the surface and low levels at the bottom, is an indication of large algal blooms. This gradient develops as a result of bacterial degradation of the elevated algal biomass that sinks to the lake sediments and high algal photosynthetic activity in the euphotic zone.

8.14 FINAL POINTS

Although anoxic sediments may not directly affect phytoplankton, they do physiologically impact the associated macrophyte root and rhizome systems(Bornette & Puijalon, 2011). These plants exhibit several adaptations to the anoxic environment, such as high reserves of materials, which enable them to use inefficient anaerobic metabolism to consume large amounts of carbohydrates. Macrophytes usually have aerenchymatous tissues that conduct air, which enables them to carry oxygen into anaerobic environments. Vascular plants that live in aquatic environments have developed adaptations that allow them to withstand extended periods of anoxic stress. These adaptations are especially well-developed in macrophytes, or heterophytes, which are rooted in low-oxygen sediments near the water surface. In both mesotrophic and oligotrophic environments, the primary factor limiting both types of organisms is the lack of mineral nutrients. Rooted vascular plants are present with increased nutrient levels and do have some advantages over both planktonic and attached filamentous algae in limiting nutrient conditions, even though oligotrophic water bodies tend to show slight development of macrophytes. While algae have limited chance of growing in the low-nutrient water column, these plants can absorb scarce mineral nutrients through their root systems and recycle them internally. Because of the abundance of nutrients in the water column and sediments, algae can grow rapidly and outcompete macrophytes in eutrophic environments. Increased phosphorus release from anoxic sediments under such circumstances promotes algal growth. Phosphorus is then carried upward through the water column by thermal mixing that occurs during the night.

As the wavelength of the irradiation is reduced, the harmful effects of light grow. Moving towards the ultraviolet end of the spectrum, which includes UV-A (315-400 nm) and UV-B (280-315 nm) light, significantly increases the inhibitory effects observed with photosynthetically active radiation (PAR 400-700 nm). UV-B content of light incident at the water surface varies based on solar angle and ozone layer thickness. When UV radiation penetrates the water column, it is rapidly lost; in clear water, 99 percent of UV-A radiation is absorbed between 2 and 5 meters below the surface, while UV-B radiation is absorbed between 0.5 and 1 meter below the surface. UV-B light will have a substantial impact on microorganisms in surface waters despite the water column attenuation of this radiation. Consequently, many researchers have employed this wavelength band in their investigations into the molecular mechanism of photoinhibition. With related microbiological impacts, incident UV-B levels are anticipated to rise in tandem with anticipated drops in the stratospheric ozone layer.

Chapter 9

MICROBES IN BIOREMEDIATION

Contents

9.1 INTRODUCTION

Bioremediation is a biological process that uses living organisms to clean up contaminated soil and water, and microbes are essential to it. It is possible to use bacteria, fungi, algae, etc. for bioremediation.

Recalcitrant chemicals in soils and sediments may be biodegraded by microorganisms to a residual concentration that either slowly drops over years or no longer decreases with time. The availability of the resistant chemicals to microbes is thought to be a limiting factor. The outcomes of using surfactants to try to boost this availability have been inconsistent. Residual recalcitrant substances may be considerably less hazardous and less leachable by water as a result of binding with soils and sediments and slow-release rates. Pollutants can only be broken down when they are present in the aqueous phase and come into contact with cell membranes. The contamination acts as a substrate for the bacterium, enters the cell by membrane transport, and fuels the metabolic processes of the cell. Nonetheless, the contamination bulk solution concentration may be restricted by physical or chemical processes, which will severely hinder the microbial capacity to absorb the contaminant. Thus, the total biodegradation of these compounds can be regulated by the bioavailability of the pollutant. Microorganism density plays a significant role in biodegradation and bioavailability(Marschner & Kalbitz, 2003). Most bacteria found in the environment are attached to the surface, and they are sporadically distributed across soils and sediments. Most of these bacteria are between 0.5 and 1.0 μm in size, while the micropores found in soils and sediments are much smaller than 1 μm. Researchers agree that bacteria are primarily connected to the outside/external surfaces of soils and sediments rather than the inside surfaces of micropores. The microbial–contaminant interactions take place in the biofilm that forms on the surfaces of soils and sediments within macropores.

9.2 BIOAVAILABILITY

This shows that soil/sediment bioavailability is controlled by the partitioning of an organic pollutant from the solid to the aqueous phase in the larger pore spaces. Chemical bonds, surface complex forms, electrostatic interactions, and hydrophobic effects are a few examples of these partitioning mechanisms. The amount of organic matter in the soil or sediment and the degree of hydrophobicity of PAH are the two factors that influence sorption for hydrophobic pollutants, such as recalcitrant chemicals. The mass transfer of the sorbate molecules from sorption sites on and in the soil/sediment is crucial (rate of desorption). Where desorption is intense, there should be larger amounts of accessible PAHs, which should be correlated with active bacteria. The locations in which recalcitrant chemicals accumulate when they are associated with soils and sediments are currently poorly understood. Therefore, to derive sorbent/sorbate mechanistic models, researchers must rely on inferences from macroscopic experimental observations that capture overall behavior and offer empirical evidence. We can assume that no single sorption mechanism predominates. Indeed, rate-limited sorption behavior is probably the result of many processes. The current techniques for evaluating the sorption and sequestration of recalcitrant compounds on soils and sediments do not offer information to help interpret the results of ecotoxicological testing of residuals after biotreatment, nor do they offer a fundamental understanding of what can be achieved by biostabilization or the bioavailability of recalcitrant compounds. It is necessary to comprehend the mechanisms that bind contaminated refractory compounds within soil or sediment in order to determine whether residual recalcitrant compounds that remain after biotreatment reflect an acceptable level. The nature of PAH's association with soils and sediments is related to biostabilization.

The development of technology for boosting natural recovery processes involved in the in situ biotreatment of soils and sediments contaminated with recalcitrant compounds, as well as research into the factors influencing the bioavailability of recalcitrant compounds on soils and sediments, are both necessary. Enhancing our knowledge of the mechanisms underlying the sequestration and bioavailability of recalcitrant chemicals in soils and sediments requires attention. The growth cycle of the microbial population can be broken down into the lag phase, exponential phase, stationary phase, and death phase. The lag phase is characterized by a delay in the expansion of the microbial population until the bacteria have adapted to the environment and the food source. Before the microorganisms have the essential metabolites and enzymes to break down the contaminant, they cannot absorb food. The bacteria go into the exponential phase of growth after producing the required enzymes and metabolites. Both the characteristics of the organism and the environment, such as temperature, have an impact on the pace of exponential growth. On the other hand, exponential growth is not sustainable. Usually, exponential growth stops when an organism's waste products accumulate to an inhibitory level or when a critical nutrient for growth is depleted. The population is now in the stationary phase, during which there is no net change in the number of cells. The microbial population will enter the death phase and the viable number of microorganisms will decline if conditions worsen. This happens if toxins continue to accumulate or the food source is exhausted. Almost all organic compounds can degrade given this growth cycle, given the right conditions or amount of time. Nevertheless, bioremediation may be hampered by a variety of materials or physical, chemical, and biological factors(Boopathy, 2000). The treatment objectives must identify the biodegradability of wastes in terms of a reasonable time frame. The limits of biodegradability resulting from toxicity are contingent upon concentration. Many toxic wastes or constituents can biodegrade if the biomass is large enough or the toxicant concentration can be regulated. Microbes can be forced to break down more waste by gradually raising the concentrations of contaminants. The logic is to gradually raise the concentrations of contaminants so that the population of microbes can adjust to the changing environment and generate the necessary metabolites and enzymes.

9.3 METALS AND OTHER CONDITIONS

Metals can impede several biological functions, and their effects are frequently concentration-dependent. Specific chemical reactions are typically involved in metal toxicity for microorganisms. While lead, barium, and iron are generally safe for bacteria to handle at normal levels, metals like copper, silver, and mercury are usually highly hazardous, especially when present as ions. In fertile soils, the nutritional metals that plants and microorganisms require are typically present in naturally occurring quantities. The main inorganic nutrients are nitrogen and phosphorus, but for the best possible biological growth, trace amounts of potassium, calcium, sulfur, magnesium, iron, and manganese are also needed. These metal toxicity or availability to microorganisms typically depends on pH, with lower pH values making the metals more mobile or available. Certain plant species and microbes have the ability to actively absorb metals. Living cells have the ability to concentrate inorganic substances and absorb metals. Heavy metals are absorbed by the biomass as a byproduct of the cell's regular metabolic activity, even though they may not be biologically necessary.

Through several processes, including ion exchange at the cell walls, complexation reactions at the cell walls, and intra- and extracellular complexation reactions, activated biomass extracts metals from solution. Metals are absorbed by inactivated biomass by adsorbing them to ionic groups on the cell surface or in the polysaccharide coating that most bacteria have. The metals are bonded to polymers either by sorption or by the exchange of functional groups. The majority of bacteria grow in a temperature range between around 10 and 38°C(Marchant et al., 2002). Up to a certain temperature, the pace of biochemical reactions in cells increases; beyond this point, the rate of activity decreases due to enzyme denaturation, which results in the death or reduced activity of organisms. The temperature has an impact on gas solubilities as well, therefore it needs to be considered while creating a remediation system. On the other hand, the temperature of ex situ activities can only be moderated, often at a significant cost, while the temperature of in situ processes is very difficult to control.

The effects of the predicted temperatures must be taken into consideration. Excessive water saturation can cause anaerobic activity in the soil, which can be advantageous or detrimental depending on the preference for aerobic or anaerobic breakdown processes. A 30 to 35% saturation of the pore spaces is typically enough to allow air to travel through the subsoil if aerobic conditions are required. However, if anaerobic conditions are required, 100% pore space saturation will be more than adequate. Abrupt changes in the pH of the waste-treatment system matrix and extreme pH values can interfere with microbial metabolism, gas solubility in soil water, nutrient availability in soil water, and heavy metal solubilities, all of which can hinder microbial growth. The pH range of most natural settings, or soils, is 5.0-9.0. As a result, this pH range is ideal for waste biodegradation boosted by microbes. Because of the presence of carbonates and other minerals, most fertile soils naturally have a buffering capacity that keeps the pH range stable. Microbes can adapt to a wider range of pH levels, but growth and metabolic rates usually decline. The diversity of microbial strains also decreases, and the bacteria become more specialized to survive in specific environments. Excessive concentrations of certain pollutants in the aqueous phase can cause problems, much as low concentrations of contaminants might make bioremediation more difficult. Certain substances are hazardous to microorganisms at high concentrations, even though they break down quickly at lower amounts. Toxicants inhibit or stop metabolic processes and frequently stop the development of new biomass, which is necessary to promote the elimination of contaminants. The concentration of toxicants, the microorganisms exposed, and their specific processes all affect the degree of toxicity. When at least one of the crucial phases in the multiple physiological processes that microbial cells go through is inhibited, the cells stop working.

9.4 BIODEGRADATION

Certain organic compounds are toxic to specific living forms, including plants and insects, and they may even be toxic to microorganisms. Herbicides, insecticides, rodenticides, fungicides, and pesticides are some of these substances. Many bacteria degrade inorganic chemicals, such as cyanides and azides; yet, these compounds may undergo degradation after a period of microbial adaptation. The mass transfer rates of nutrients, pollutants, water, air (i.e., oxygen), and pH adjusters are all influenced by the type of soil. The functioning of the degradation process and the possibility of waste movement are impacted by mass transport. Organic soils have the potential to be sorptive and block the flow of organic matter. The unsaturated and saturated treatment zones need to be taken into account when understanding and planning a bioremediation system. The unsaturated zone is made up of solid particles with different sizes that are related to both air and water. The lithology, recharge, and discharge characteristics of the site determine how much air and water are contained in these pore spaces. Under the effect of gravity, the water in these pore spaces gradually percolates until it reaches the saturated zone. The saturated zone is the area where liquid water makes up 100% of the pore volume. Groundwater will be produced in this area by formations that hold water. An aquifer is defined as groundwater that is present in adequate quantities. Chemical, biological, and physical interactions can be confined, in contrast to the unsaturated zone(Nielsen et al., 1986). The saturated zone is the most challenging area for biotreatment because of this property. The saturated and unsaturated zones are demarcated by the water table. Both zones exhibit specific patterns of pollution. A portion of the contamination that is retained by capillary forces often remains when the contamination drains through the unsaturated zone. The soil air in the vadose zone is filled with a plume of vapors if the pollutant is volatile. Certain pollutants seep or penetrate deeper into the water table before lateral dispersion. As groundwater flows through the aquifer, it encounters the pollutants and carries the substances that are soluble in water.

9.5 CHLORINATED COMPOUNDS

Only compounds with fewer than five chlorine substituents can be broken down by aerobic bacteria. Anaerobic bacteria remove one or more chlorine atoms during the dechlorination process and replace them with hydrogen ions to create a less hazardous, less carcinogenic, and more easily biodegradable chemical. Chemical and mechanical methods can be used for dechlorination. Chlorinated compounds become readily degradable and lose chlorine ions when they are combined with metallic calcium in an ethanol solvent at ambient temperature and atmospheric pressure. This procedure has the advantage of not requiring high pressures or temperatures. This approach is still in the experimental stage and is not yet suitable for application. The problem with this process is that dechlorination does not fulfill regulatory standards for pollutant elimination. Although toxicity reduction renders polluted soils as safe as other treatments, these requirements frequently focus on mass removal rather than toxicity reduction. Therefore, dechlorination needs to be used in conjunction with another remediation method. Although compost has traditionally been used to break down yard waste, food waste, and sewage, it is also being used to break down explosives and PAHs, among other toxins. Contaminated soil is combined with organic materials like corn cobs, activated sludge, and cow dung. The biomass required for worm and microbial growth is supplied by these additions. The contaminant is eliminated by the bacteria while the worms aerate the soil. The compost naturally experiences phases of microbial growth and decline, allowing the system to self-regulate. The only human intervention needed after the compost and contaminated soil are combined is the occasional mixing. The drawback is that complex pollutants, like molecules with five or six rings, cannot be removed using this technique.

Bench level analysis concentrates on specific elements. It is possible to quantify the relative effects of different parameters such as the type and concentration of contaminants, the species of microbe, etc. on the overall process by controlling them. The purpose of these bench tests is to collect data on the degradation processes so that full-scale site remediation solutions can make use of that information. However, the diverse soil and variety of pollutants seen in many field sites may change the rates of degradation.

Because of these competing variables, scaling up operations from the bench level to the field level is challenging. For energy, most bacteria and humans oxidize organic molecules(Imlay, 2003). This process involves the transport of electrons or their reducing equivalents from the organic component, also known as the substrate or electron donor to an electron acceptor, where they are eventually reduced. A chemical that is inorganic or organic could be the electron acceptor. Through a complicated series of oxidation-reduction (redox) reactions and the creation of energy storage compounds or electrochemical gradients, useful energy is recovered during this electron transfer process.

Aerobic heterotrophic respiration is the process of oxidizing organic molecules while reducing molecular oxygen. Anaerobic respiration is the process by which bacteria link the oxidation of organic substances to alternative electron acceptors, such as SO_4, nitrate, ferric iron, and chloroethenes. Oxidative catabolism is the process of oxidizing organic materials in conjunction with oxygen reduction or other electron acceptors to produce energy and carbon molecules needed for microbial development. Environments in oxygenated groundwater are limited in electron donors.

In the subsurface, easily degradable organic matter can create a range of geographically and chronologically distinct metabolic zones when it penetrates in large amounts. These zones rely on electron acceptors. Whether organic matter is an inflow of natural material or a human-produced (anthropogenic) pollutant, when it enters an oxygenated aquifer, native aerobic heterotrophic microbes digest it and use up the available oxygen. Aerobic respiration slows down. This makes it possible for anaerobic metabolic communities to grow and make use of different electron acceptors. In these conditions, the subsurface environment becomes electron acceptor limited, and the availability of different electron acceptors controls the rates at which the injected organics convert.

9.6 CHLOROETHENES

Since the parent chloroethenes (PCE and TCE) do not provide bio-oxidizable electron donors to subsurface habitats, the affected groundwater will not become anoxic. However, chloroethenes are frequently found in other organic materials or may mobilize naturally occurring or man-made materials. These materials can act as electron donors and be used to reduce dissolved oxygen, other anaerobic electron acceptors, or chloroethenes. The amount of reducing equivalents used for the dechlorination of chloroethenes is determined by the relative concentrations of all available electron acceptors. The parent chloroethenes are not easily broken down by oxidative processes. This indicates that the biodegradation capacity of these chemicals varies from not detectable (PCE) to low (vinyl chloride (VC) in the presence of oxygen(Sing et al., 2004). Under cometabolic conditions, TCE and less chlorinated molecules can be broken down. However, the significance of this process is restricted because it depends on the existence of an inducer and is not directly beneficial, in fact, it might even be harmful to the bacteria. Microorganisms can undergo a process called reductive dechlorination to convert chloroethenes in anaerobic settings. Through this process, the chlorinated ethenes are converted to nonchlorinated compounds (ethane and ethene), which are safe for the environment and human health. When certain microbes employ halogenated chemicals, like PCE, as terminal electron acceptors to obtain energy for development, the process is known as halorespiration, or reductive dechlorination. The intermediates in the halorespiration pathway are the more mobile and hazardous products, namely dichloroethene (DCE) and vinyl chloride (VC). The less chlorinated chloroethenes may occasionally go through oxidative catabolism in anaerobic environments. In areas where redox potentials are higher, this process might play a significant role in the mass removal of vinyl chloride.

9.7 *IN SITU* TECHNIQUES

Two in situ techniques that are frequently employed for the remediation of dissolved phase pollutants include bioremediation and natural attenuation. Utilizing microbial metabolic activities to break down, detoxify, or mineralize pollutants is known as bioremediation. When it comes to contaminated subsurface habitats (soil, sediments, and groundwater), bioremediation typically entails injecting nutrients to encourage the targeted biological activities. Another name for this is biostimulation. Introducing or injecting specific microbes, typically in conjunction with nutrients, is a less popular but increasingly popular method known as bioaugmentation(Gao et al., 2022). A remediation method known as natural attenuation makes use of the commonly known fact that subterranean contamination concentrations tend to drop with time. This is caused by a variety of biological, chemical, and physical processes that break down, immobilize, or change dissolved pollutants. The most significant of these attenuation mechanisms for the majority of pollutants is natural microbial biodegradation. These technologies have relatively low treatment costs and, in most cases, result in the destruction of the target contaminant; however, because these processes are slow, they may not be practical given the risk of public exposure to contaminants that may be present in groundwater that could reach surface water or water wells. In addition, engineers do not have sufficient experience with either bioremediation or natural attenuation, and there are some uncertainties regarding long-term performance, especially about natural attenuation. Several effective bioremediation technology demonstrations for chlorinated solvent sites were documented. Nonetheless, a few restrictions were specific to chlorinated solvent sites and DNAPL. For oxidizable pollutants like petroleum, both active and passive bioremediation techniques are metabolically robust and redundant. The catabolic capacity was discovered across a range of redox states and populations, allowing for the sequential employment of different electron acceptors to promote degradation. The conditions that permitted transformation, at least for total dechlorination, were less favorable for chlorinated solvents, and this capacity does not seem to be ubiquitous. Significantly lower rates of activity and possible instability in active systems were the consequences of these constraints. The fact that the methods are focused on dissolved pollutants and have not yet been demonstrated to clean up areas affected by other phase contaminants is another possible drawback. Many studies have been carried out to show the biological changes of dissolved phase chloroethenes. Complete dechlorination can occur even at initial PCE values as high as 91 mg/l.

There has never been an attempt to combine solvent flushing and bioremediation in the field, although these technologies have been demonstrated in lab settings and in a few field assessments to improve the cleanup of contaminated aquifers. To ensure the success of implementation, the system design must take into account many important factors, including the high degree of spatial variability in the subsurface environment, the changes in the hydraulic properties as the DNAPL is removed, and the effects of high concentrations of remedial solvent mixtures on the microbial ecology. Researchers from the Environmental Protection Agency (EPA) started assessing possible cosolvents and their breakdown product impacts on the reductive dechlorination process. To achieve whole site restoration, the Solvent Extraction Residual Biotreatment (SERB) method is a treatment strategy that combines cosolvent extraction and in situ bioremediation. The location and size of the source as well as the presence of a separate phase DNAPL are determined through site characterization, but frequently these processes are not carried out to provide the level of detail necessary for the effective deployment of many remediation solutions. Gravity causes DNAPL to flow through the aquifer and accumulate on less permeable layers. The number and locations of injection and recovery wells for the cosolvent extraction process must be determined by site characterization, optimization modeling, and groundwater flow characterization. The characteristics of the contaminant to be eliminated, the cosolvent properties, and any legal restrictions on compound injection are taken into consideration when choosing a cosolvent. The improved dissolving of the contaminant into the cosolvent, which was then eliminated through the recovery/extraction wells, served as the foundation for the cosolvent extraction method used in the SERB.

9.8 OTHER TECHNOLOGIES

In addition to being removed by the recovery/extraction wells, some in situ extraction methods are made to mobilize the residual contamination. In the case of DNAPL contamination in particular, mobilization of the contaminant necessitates good hydraulic control to guarantee that the DNAPL is prevented from further contaminating the aquifer(Rivett et al., 2001). The cosolvent flushing test is carried out when the cosolvent is chosen and the injection/extraction design is established. For the technique to be successfully applied, the most crucial design criterion for the injection/extraction system is maximum contact and mixing of the cosolvent and residual contaminant. It is recognized that some DNAPL will not be contacted by the cosolvent due to the complexity of the subsurface environment, and that the majority of the flow will occur via layers with higher hydraulic conductivity.

It is anticipated that there would be distinct regions of cosolvent and DNAPL pollution in the aquifer once the cosolvent extraction method is applied and active pumping is stopped, but there will also be areas where the cosolvent and DNAPL are combined.

The regions with high potential to promote bioremediation are those when the cosolvent (electron donor) and DNAPL (electron acceptor) are combined. Since ethanol was a suitable electron donor and cosolvent for the reductive dechlorination process, it was chosen as the cosolvent. Subsurface bioactive zones created by bioremediation processes should be able to remove contaminants from groundwater more quickly than they dissolve from the residual DNAPL in the subsurface. Achieving this would allow for the full restoration of the site. The most crucial factor in the injection/extraction system design for the SERB technology is the mixing of the cosolvent and the contaminant for the separate phase DNAPL dissolution and removal, as well as the biodegradation of the residual contaminant. For many in situ treatment approaches, getting the electron donor to the subsurface has proven difficult and is typically constrained by transportation issues. The cosolvency effect and the injection/extraction mechanism enable the simultaneous exposure of the electron donor (cosolvent) to the electron acceptor (chlorinated solvent) during the cosolvent flush.

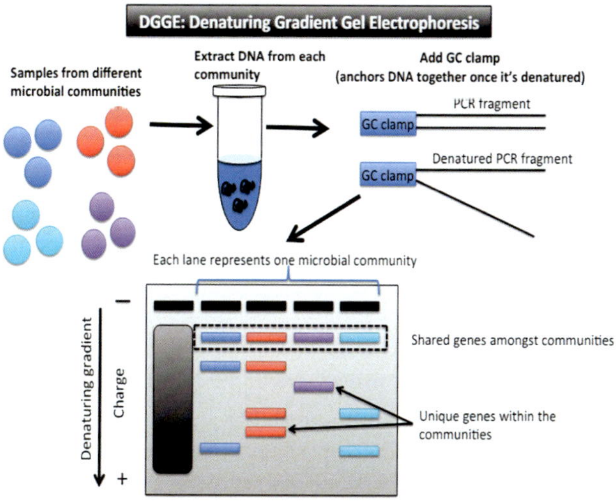

Figure 9.1 Denaturing gradient gel electrophoresis

Source: https://upload.wikimedia.org/wikipedia/en/6/68/Step-by-step_procedure_of_using_DGGE_analysis_in_microbiology.pdf,*Creative Commons License.*

The establishment of conducive conditions for microorganisms to break down the target contaminant is a prerequisite for the success of any bioremediation method. Failures in bioremediation are more frequently caused by ignorance of ecological restrictions than by understanding the organism's genetic potential. The addition or modification of genetic traits may have a negative impact on the environmental fitness of microorganisms, although a multitude of molecular markers have been developed and employed for the detection of microorganisms, including successful applications based on the green fluorescent protein (GFP) gene. A quantitative component is absent from other molecular tracking techniques, such as denaturing gradient gel electrophoresis (DGGE) (Figure 9.1), terminal restriction fragment length polymorphism (T-RFLP), single-strand conformation polymorphism (SSCP), and amplified ribosomal DNA restriction analysis (ARDRA). Quantitation can be achieved through in situ hybridization using fluorescently labeled oligonucleotide probes (FISH) that are targeted to the rRNA sequences. This method has been shown to be effective in aqueous environments, but it is labor-intensive, does not provide species-to-strain specificity, and has been challenging to apply in soil. The original purpose of real-time PCR (RT-PCR) was to measure the copy number of a specific gene. This technique was used for the identification and measurement of microorganisms. This technique uses a double-labeled probe to track the amount of fluorescence that a released reporter dye accumulates during PCR. When PCR is more quantitative, this fluorescence is then connected to the quantity of product generated in real-time during amplification. Using this method, we can measure the 16S rRNA gene and the 4-chlorobenzoate *fcb* degradation operon to investigate the population dynamics in the soil of a recombinant Rhodococcus RHA1 genetically modified to grow on a chlorobiphenyl(Rodrigues et al., 2002).

9.9 REMEDIATION APPROACHES FOR PAH

Remediation of PAH-contaminated soils through land farming is a widely used and generally dependable approach that relies on the degradative actions of natural microbial communities. It is typically chosen over nonbiological techniques like stabilization, chemical oxidation, and incineration. It has been used for a range of soil types and contaminants such as petroleum, coal tar, and creosote. By supplying oxygen and nitrogen, land farming aims to enhance the natural microbial capacity for degradation. Physical soil mixing is employed to disperse contaminants across a larger surface area and facilitate their interaction with microorganisms. Conventional farming methods are employed to supply nitrogen (fertilizer), water (irrigation), and mixing (tilling). Fertilizers can be used as organic waste (manure). Land farming is commonly, though not always, restricted to treating 6 to 12 inches of soil at a time (the depth of tilling); however, after one layer (lift) is well treated, other lifts can be applied on top. The time and operation expenses, and space needed for the lifts are the drawbacks of traditional land farming. Solid phase remediation frequently results in a two-stage process of degradation: a rapid phase that lasts for several months and involves extensive PAH degradation, and a second phase that takes months to years and involves relatively minimal change in PAH concentration. The two-stage process is commonly attributed to variations in the speeds at which soil particles desorb into the interstitial water. Desorption proceeds quickly, and microbiological variables control the rate of degradation. Both low and high-molecular-weight PAHs can be partially broken down by a variety of fungal species. Fungal lignolytic enzymes, such as laccases and peroxidases, are typically responsible for the early reactions of PAH degradation. These organisms may also play a role in the turnover of PAHs in uncontaminated soils. The efficiency of a coculture of the bacterium Stenotrophomonas, a mixed bacterial population, and a fungus (Penicillium) in breaking down five-ring PAHs implies that the bacteria subsequently break down the first oxidation products that the fungi produce. There was no proof that this coculture had an impact on the HMW PAH bioavailability. Although it might not be a very helpful method for bioremediation, it is interesting to consider how naturally occurring degradation of PAHs in the environment could occur.

Stenotrophomonas maltophilia degraded mixtures of PAHs more quickly in the presence of surfactant, and that specific degradation rate from the mixture was lower when degraded in combination than when degraded separately(González et al., 2011). Similar drops in the rate of individual PAH degradation brought on by the presence of other PAHs have been seen for pseudomonads and mixed cultures in the absence of surfactants. More research in this field is required to improve our knowledge of how surfactants affect mixed microbial decomposition and to standardize surfactant application as an engineering tool for biological soil remediation. *Sphingomonas paucimobilis* degraded mixtures of PAHs while Tween 80 was present. Despite their increased availability to the microbes, phenanthrene, fluoranthene, and pyrene degrade in a mixture in the presence of Tween 80. The preference for phenanthrene over fluoranthene and fluoranthene over pyrene for this strain is likely due to the selectivity of the dioxygenase system in the cells. When microbes are cultured on insoluble or immiscible chemicals, they produce surface-active agents known as biosurfactants. To increase the bioavailability of hydrophobic pollutants, the use of biosurfactants has been suggested as an alternative to chemical surfactants. A structurally varied class of surface-active chemicals is known as biosurfactants. They are amphipathic compounds, typically in the form of fatty acids, with a hydrophilic part and a hydrophobic moiety. They have low values of critical micelle concentration (CMC) and interfacial tension (IFT).

The quantity of surfactant required to reach the lowest feasible surface tension is known as the CMC. CMC is the surfactant concentration at which micellar structures will form with the addition of free monomer. Slightly soluble substances can be scattered into the aqueous phase by the aggregating surfactant molecules during the micelle formation process. Compared to chemical surfactants, biosurfactants have a CMC that is roughly ten to forty times lower. Based on their general structure, biosurfactants can be categorized into three groups: high-molecular-weight biopolymers, lipopeptides, and glycolipids. Various sugar moieties, including rhamnose and trehalose, are joined to long-chain fatty acids to form glycolipids. A short polypeptide joined to a lipid moiety is what makes up a lipopeptide. Lipoprotein, lipopolysaccharide-protein complexes, and polysaccharide-protein-fatty acid complexes are examples of high-molecular-weight biopolymers. Many bacteria, such as Pseudomonas sp., Bacillus sp., and Mycobacterium sp., are known to create biosurfactants.

REFERENCES

1. Aarsman, M. E. G., Piette, A., Fraipont, C., Vinkenvleugel, T. M. F., Nguyen-Distèche, M., & den Blaauwen, T. (2005). Maturation of the Escherichia coli divisome occurs in two steps. *Molecular Microbiology*, *55*(6), 1631–1645. https://doi.org/10.1111/j.1365-2958.2005.04502.x

2. Abeijón Mukdsi, M. C., Medina, R. B., Alvarez, M. D. F., & González, S. N. (2009). Ester synthesis by lactic acid bacteria isolated from goat's and ewe's milk and cheeses. *Food Chemistry*, *117*(2), 241–247. https://doi.org/10.1016/j.foodchem.2009.03.105

3. Addinall, S. G., & Holland, B. (2002). The Tubulin Ancester, FtsZ, Draughtsman, Designer and Driving Force for Bacterial Cytokinesis. *Journal of Molecular Biology*, *318*(2), 219–236. https://doi.org/10.1016/S0022-2836(02)00024-4

4. Addinall, S. G., & Lutkenhaus, J. (1996). FtsZ-spirals and -arcs determine the shape of the invaginating septa in some mutants of Escherichia coli. *Molecular Microbiology*, *22*(2), 231–237. https://doi.org/10.1046/j.1365-2958.1996.00100.x

5. Ahmad, M., Nadeem, S. M., & Zahir, Z. A. (2019). Plant-Microbiome Interactions in Agroecosystem: An Application. In V. Kumar, R. Prasad, M. Kumar, & D. K. Choudhary (Eds.), *Microbiome in Plant Health and Disease: Challenges and Opportunities* (pp. 251–291). Springer. https://doi.org/10.1007/978-981-13-8495-0_12

6. Aldrian, D., Waldner, B., Vogel, G. F., El-Gharbawy, A. H., McKiernan, P., Vockley, J., Landau, Y. E., Al Mutairi, F., Stepien, K. M., Kwok, A. M.-K.,

Yıldız, Y., Honzik, T., Kelifova, S., Ellaway, C., Lund, A. M., Mori, M., Grünert, S. C., Scholl-Bürgi, S., Zöggeler, T., … Karall, D. (2024). Impact of citrulline substitution on clinical outcome after liver transplantation in carbamoyl phosphate synthetase 1 and ornithine transcarbamylase deficiency. *Journal of Inherited Metabolic Disease*, *47*(2), 220–229. https://doi.org/10.1002/jimd.12717

7. Allakhverdiev, S. I. (Ed.). (2015). *Photosynthesis: New Approaches to the Molecular, Cellular, and Organismal Levels* (1st ed.). Wiley. https://doi.org/10.1002/9781119084150
8. Amballa, H., & Bhumi, N. R. (2016). Significance of Arbuscular Mycorrhizal Fungi and Rhizosphere Microflora in Plant Growth and Nutrition. In D. K. Choudhary, A. Varma, & N. Tuteja (Eds.), *Plant-Microbe Interaction: An Approach to Sustainable Agriculture* (pp. 417–452). Springer Singapore. https://doi.org/10.1007/978-981-10-2854-0_19
9. Antelmann, H., Williams, R. C., Miethke, M., Wipat, A., Albrecht, D., Harwood, C. R., & Hecker, M. (2005). The extracellular and cytoplasmic proteomes of the non-virulent Bacillus anthracis strain UM23C1-2. *PROTEOMICS*, *5*(14), 3684–3695. https://doi.org/10.1002/pmic.200401218
10. Arnell, N. W., Cannell, M. G. R., Hulme, M., Kovats, R. S., Mitchell, J. F. B., Nicholls, R. J., Parry, M. L., Livermore, M. T. J., & White, A. (2002). The Consequences of CO2 Stabilisation for the Impacts of Climate Change. *Climatic Change*, *53*(4), 413–446. https://doi.org/10.1023/A:1015277014327
11. Balleza, D., & Gómez-Lagunas, F. (2009). Conserved motifs in mechanosensitive channels MscL and MscS. *European Biophysics Journal*, *38*(7), 1013–1027. https://doi.org/10.1007/s00249-009-0460-y
12. Becher dos Passos, J., Vanhalewyn, M., Lopes Brandão, R., Castro, I. M., Nicoli, J. R., & Thevelein, J. M. (1992). Glucose-induced activation of plasma membrane H+-ATPase in mutants of the yeast *Saccharomyces cerevisiae* affected in cAMP metabolism, cAMP-dependent protein phosphorylation and the initiation of glycolysis. *Biochimica et Biophysica Acta (BBA) - Molecular Cell Research*, *1136*(1), 57–67. https://doi.org/10.1016/0167-4889(92)90085-P
13. Bennett, R. K., Steinberg, L. M., Chen, W., & Papoutsakis, E. T. (2018). Engineering the bioconversion of methane and methanol to fuels and chemicals in native and synthetic methylotrophs. *Current Opinion in Biotechnology*, *50*, 81–93. https://doi.org/10.1016/j.copbio.2017.11.010
14. Berg, G., Roskot, N., Steidle, A., Eberl, L., Zock, A., & Smalla, K. (2002).

Plant-Dependent Genotypic and Phenotypic Diversity of Antagonistic Rhizobacteria Isolated from Different *Verticillium* Host Plants. *Applied and Environmental Microbiology*, *68*(7), 3328–3338. https://doi.org/10.1128/AEM.68.7.3328-3338.2002

15. Bhatla, S. C., & Lal, M. A. (2023). Nitrogen Metabolism. In S. C. Bhatla & M. A. Lal (Eds.), *Plant Physiology, Development and Metabolism* (pp. 295–334). Springer Nature. https://doi.org/10.1007/978-981-99-5736-1_11
16. Bicas, J. L., & Rodriguez-Amaya, D. B. (2021). Generation of process-derived flavors and off-flavors. In *Chemical Changes During Processing and Storage of Foods* (pp. 385–451). Elsevier. https://doi.org/10.1016/B978-0-12-817380-0.00009-9
17. Blum, J. L. (1957). An ecological study of the algae of the Saline River, Michigan. *Hydrobiologia*, *9*(4), 361–408. https://doi.org/10.1007/BF00150384
18. Bonaldo, F., & Leroy, F. (2024). Review: Bacterially produced vitamin K2 and its potential to generate health benefits in humans. *Trends in Food Science & Technology*, *147*, 104461. https://doi.org/10.1016/j.tifs.2024.104461
19. Boopathy, R. (2000). Factors limiting bioremediation technologies. *Bioresource Technology*, *74*(1), 63–67. https://doi.org/10.1016/S0960-8524(99)00144-3
20. Borbély, G., Surányi, G., & Kós, P. (1990). Stress responses of cyanobacteria and the pleiotropic effects of light deprivation. *FEMS Microbiology Letters*, *74*(2), 141–152. https://doi.org/10.1016/0378-1097(90)90549-6
21. Bornette, G., & Puijalon, S. (2011). Response of aquatic plants to abiotic factors: A review. *Aquatic Sciences*, *73*(1), 1–14. https://doi.org/10.1007/s00027-010-0162-7
22. Boron, W. F. (2004). Regulation of intracellular pH. *Advances in Physiology Education*, *28*(4), 160–179. https://doi.org/10.1152/advan.00045.2004
23. Cabeen, M. T., Herrmann, H., & Jacobs-Wagner, C. (2011). The domain organization of the bacterial intermediate filament-like protein crescentin is important for assembly and function. *Cytoskeleton*, *68*(4), 205–219. https://doi.org/10.1002/cm.20505
24. Camacho-Chab, J. C., Lango-Reynoso, F., Castañeda-Chávez, M. D. R., Galaviz-Villa, I., Hinojosa-Garro, D., & Ortega-Morales, B. O. (2016). Implications of Extracellular Polymeric Substance Matrices of Microbial Habitats Associated with Coastal Aquaculture Systems. *Water*, *8*(9), Article 9. https://doi.org/10.3390/w8090369

25. Carballido-López, R. (2006). The Bacterial Actin-Like Cytoskeleton. *Microbiology and Molecular Biology Reviews*, *70*(4), 888–909. https://doi.org/10.1128/MMBR.00014-06
26. Cesselin, B., Derré-Bobillot, A., Fernandez, A., Lamberet, G., Lechardeur, D., Yamamoto, Y., Pedersen, M. B., Garrigues, C., Gruss, A., & Gaudu, P. (2011). Responses of Lactic Acid Bacteria to Oxidative Stress. In E. Tsakalidou & K. Papadimitriou (Eds.), *Stress Responses of Lactic Acid Bacteria* (pp. 111–127). Springer US. https://doi.org/10.1007/978-0-387-92771-8_6
27. Chamberlin, M. J. (1982). 3 Bacterial DNA-Dependent RNA Polymerases. In P. D. Boyer (Ed.), *The Enzymes* (Vol. 15, pp. 61–86). Academic Press. https://doi.org/10.1016/S1874-6047(08)60275-9
28. Chen, A., Wang, D., Nugen, S. R., & Chen, J. (2021). An Engineered Reporter Phage for the Fluorometric Detection of Escherichia coli in Ground Beef. *Microorganisms*, *9*(2), Article 2. https://doi.org/10.3390/microorganisms9020436
29. Chen, H. (2014). Chemical Composition and Structure of Natural Lignocellulose. In H. Chen (Ed.), *Biotechnology of Lignocellulose: Theory and Practice* (pp. 25–71). Springer Netherlands. https://doi.org/10.1007/978-94-007-6898-7_2
30. Chen, W., & Lu, W. (2018). Environmental Stress Responses of Lactic Acid Bacteria. In W. Chen & A. Narbad (Eds.), *Lactic Acid Bacteria in Foodborne Hazards Reduction: Physiology to Practice* (pp. 115–140). Springer. https://doi.org/10.1007/978-981-13-1559-6_5
31. Cheng, K. Y., Mohottige, T. W., Ginige, M. P., & Kaksonen, A. H. (2023). New perspectives for bio-technical treatment of oxalate-containing waste streams from bauxite processing. *Hydrometallurgy*, *220*, 106105. https://doi.org/10.1016/j.hydromet.2023.106105
32. Chet, I., & Mitchell, R. (1976). Ecological aspects of microbial chemotactic behavior. *Annual Review of Microbiology*, *30*(Volume 30,), 221–239. https://doi.org/10.1146/annurev.mi.30.100176.001253
33. Cocking, E. C. (2003). Endophytic colonization of plant roots by nitrogen-fixing bacteria. *Plant and Soil*, *252*(1), 169–175. https://doi.org/10.1023/A:1024106605806
34. Cosgrove, D. J., & Hedrich, R. (1991). Stretch-activated chloride, potassium, and calcium channels coexisting in plasma membranes of guard cells of Vicia faba L. *Planta*, *186*(1), 143–153. https://doi.org/10.1007/BF00201510

35. Das, B. (2022). Origin of Life on Earth. In *The History and Philosophy of Science*. Routledge India.
36. Dignam, B. E. A., O'Callaghan, M., Condron, L. M., Raaijmakers, J. M., Kowalchuk, G. A., & Wakelin, S. A. (2016). Challenges and opportunities in harnessing soil disease suppressiveness for sustainable pasture production. *Soil Biology and Biochemistry*, *95*, 100–111. https://doi.org/10.1016/j.soilbio.2015.12.006
37. Dillard, B. D., Demick, J. M., Adams, M. W. W., & Lanzilotta, W. N. (2011). A cryo-crystallographic time course for peroxide reduction by rubrerythrin from Pyrococcus furiosus. *JBIC Journal of Biological Inorganic Chemistry*, *16*(6), 949–959. https://doi.org/10.1007/s00775-011-0795-6
38. Dimroth, P., & Hilbi, H. (1997). Enzymic and genetic basis for bacterial growth on malonate. *Molecular Microbiology*, *25*(1), 3–10. https://doi.org/10.1046/j.1365-2958.1997.4611824.x
39. Ding, J., Xu, K., Xu, H., Ji, J., Qian, Y., & Shen, J. (2024). Advances in Gas Therapeutics for Wound Healing: Mechanisms, Delivery Materials, and Prospects. *Small Structures*, *5*(1), 2300151. https://doi.org/10.1002/sstr.202300151
40. Egorov, V. (2021). Theory of Radioisotope and Chemical Homeostasis of Marine Ecosystems. In V. Egorov (Ed.), *Theory of Radioisotopic and Chemical Homeostasis of Marine Ecosystems* (pp. 163–305). Springer International Publishing. https://doi.org/10.1007/978-3-030-80579-1_4
41. Ferenci, T. (2007). Bacterial Physiology, Regulation and Mutational Adaptation in a Chemostat Environment. In R. K. Poole (Ed.), *Advances in Microbial Physiology* (Vol. 53, pp. 169–315). Academic Press. https://doi.org/10.1016/S0065-2911(07)53003-1
42. Forrest, K. L., & Bhave, M. (2007). Major intrinsic proteins (MIPs) in plants: A complex gene family with major impacts on plant phenotype. *Functional & Integrative Genomics*, *7*(4), 263–289. https://doi.org/10.1007/s10142-007-0049-4
43. Franche, C., Lindström, K., & Elmerich, C. (2009). Nitrogen-fixing bacteria associated with leguminous and non-leguminous plants. *Plant and Soil*, *321*(1), 35–59. https://doi.org/10.1007/s11104-008-9833-8
44. Gao, D., Zhao, H., Wang, L., Li, Y., Tang, T., Bai, Y., & Liang, H. (2022). Current and emerging trends in bioaugmentation of organic contaminated soils: A review. *Journal of Environmental Management*, *320*, 115799. https://doi.org/10.1016/j.jenvman.2022.115799
45. Geck, P., & Heinz, E. (1981). Chapter 11 Coupled transport of metabolites.

In S. L. Bonting & J. J. H. H. M. de Pont (Eds.), *New Comprehensive Biochemistry* (Vol. 2, pp. 285–310). Elsevier. https://doi.org/10.1016/S0167-7306(08)60039-3

46. González, N., Simarro, R., Molina, M. C., Bautista, L. F., Delgado, L., & Villa, J. A. (2011). Effect of surfactants on PAH biodegradation by a bacterial consortium and on the dynamics of the bacterial community during the process. *Bioresource Technology*, *102*(20), 9438–9446. https://doi.org/10.1016/j.biortech.2011.07.066
47. Gowdridge, B. M. (1956). Heterocaryons between Strains of Neurospora Crassa with Different Cytoplasms. *Genetics*, *41*(5), 780–789. https://www.ncbi.nlm.nih.gov/pmc/articles/PMC1209816/
48. Graumann, P. L. (2001). SMC proteins in bacteria: Condensation motors for chromosome segregation? *Biochimie*, *83*(1), 53–59. https://doi.org/10.1016/S0300-9084(00)01218-9
49. Griggs, D. W., & Johnston, M. (1993). Promoter Elements Determining Weak Expression of the GAL4 Regulatory Gene of Saccharomyces cerevisiae. *Molecular and Cellular Biology*, *13*(8), 4999–5009. https://doi.org/10.1128/mcb.13.8.4999-5009.1993
50. Hadibarata, T., & Kristanti, R. A. (2013). Biodegradation and metabolite transformation of pyrene by basidiomycetes fungal isolate Armillaria sp. F022. *Bioprocess and Biosystems Engineering*, *36*(4), 461–468. https://doi.org/10.1007/s00449-012-0803-4
51. Hartman, K., Van Der Heijden, M. G., Roussely-Provent, V., Walser, J.-C., & Schlaeppi, K. (2017). Deciphering composition and function of the root microbiome of a legume plant. *Microbiome*, *5*(1), 2. https://doi.org/10.1186/s40168-016-0220-z
52. Hawkes, C. V., DeAngelis, K. M., & Firestone, M. K. (2007). Root Interactions with Soil Microbial Communities and Processes. In *The Rhizosphere* (pp. 1–29). Elsevier. https://doi.org/10.1016/B978-012088775-0/50003-3
53. Heichlinger, A., Ammelburg, M., Kleinschnitz, E.-M., Latus, A., Maldener, I., Flärdh, K., Wohlleben, W., & Muth, G. (2011). The MreB-Like Protein Mbl of *Streptomyces coelicolor* A3(2) Depends on MreB for Proper Localization and Contributes to Spore Wall Synthesis. *Journal of Bacteriology*, *193*(7), 1533–1542. https://doi.org/10.1128/JB.01100-10
54. Heinonen, S.-M., Wähälä, K., & Adlercreutz, H. (2002). Metabolism of isoflavones in human subjects. *Phytochemistry Reviews*, *1*(2), 175–182. https://doi.org/10.1023/A:1022555216303

55. Hou, C. T. (1984). Microbiology and Biochemistry of Methylotrophic Bacteria. In *Methylotrophs: Microbiology. Biochemistry and Genetics*. CRC Press.

56. Huemer, M., & Baumgartner, M. R. (2019). The clinical presentation of cobalamin-related disorders: From acquired deficiencies to inborn errors of absorption and intracellular pathways. *Journal of Inherited Metabolic Disease*, *42*(4), 686–705. https://doi.org/10.1002/jimd.12012

57. Hussain, S. S., Mehnaz, S., & Siddique, K. H. M. (2018). Harnessing the Plant Microbiome for Improved Abiotic Stress Tolerance. In D. Egamberdieva & P. Ahmad (Eds.), *Plant Microbiome: Stress Response* (pp. 21–43). Springer. https://doi.org/10.1007/978-981-10-5514-0_2

58. Imlay, J. A. (2003). Pathways of Oxidative Damage. *Annual Review of Microbiology*, *57*(Volume 57, 2003), 395–418. https://doi.org/10.1146/annurev.micro.57.030502.090938

59. Isaac, S., Gokhale, A. V., & Wyatt, A. M. (1986). Selectivity to K+ and Na+ of Protoplast Fractions Isolated from Different Regions of Aspergillus nidulans Hyphae. *Microbiology*, *132*(5), 1173–1179. https://doi.org/10.1099/00221287-132-5-1173

60. Izaguirre-Mayoral, M. L., Lazarovits, G., & Baral, B. (2018). Ureide metabolism in plant-associated bacteria: Purine plant-bacteria interactive scenarios under nitrogen deficiency. *Plant and Soil*, *428*(1–2), 1–34. https://doi.org/10.1007/s11104-018-3674-x

61. Jacobsen, S.-E., Jensen, C. R., & Liu, F. (2012). Improving crop production in the arid Mediterranean climate. *Field Crops Research*, *128*, 34–47. https://doi.org/10.1016/j.fcr.2011.12.001

62. Jurtshuk, P., Mueller, T. J., & Acord, W. C. (1975). Bacterial Terminal Oxidases. *CRC Critical Reviews in Microbiology*, *3*(4), 399–468. https://doi.org/10.3109/10408417509108757

63. Kagan, Z. S., Dorozhko, A. I., Kovaleva, S. V., & Yakovleva, L. I. (1975). Studies of homogeneous "biosynthetic" l-threonine dehydratase from *Escheria coli* K-12. *Biochimica et Biophysica Acta (BBA) - Enzymology*, *403*(1), 208–220. https://doi.org/10.1016/0005-2744(75)90023-6

64. Kang, C.-H., Han, S. H., Kim, J.-S., Kim, Y., Jeong, Y., Park, H. M., & Paek, N.-S. (2019). Inhibition of Nitric Oxide Production, Oxidative Stress Prevention, and Probiotic Activity of Lactic Acid Bacteria Isolated from the Human Vagina and Fermented Food. *Microorganisms*, *7*(4), Article 4. https://doi.org/10.3390/microorganisms7040109

65. Kapoor, R., & Mukerji, K. G. (2006). Rhizosphere Microbial Community Dynamics. In K. G. Mukerji, C. Manoharachary, & J. Singh (Eds.),

Microbial Activity in the Rhizoshere (Vol. 7, pp. 55–69). Springer-Verlag. https://doi.org/10.1007/3-540-29420-1_4

66. Kawasaki, A., Watson, E. R., & Kertesz, M. A. (2012). Indirect effects of polycyclic aromatic hydrocarbon contamination on microbial communities in legume and grass rhizospheres. *Plant and Soil*, *358*(1), 169–182. https://doi.org/10.1007/s11104-011-1089-z
67. Kelemen, G. H. (2017). Intermediate Filaments Supporting Cell Shape and Growth in Bacteria. In J. Löwe & L. A. Amos (Eds.), *Prokaryotic Cytoskeletons: Filamentous Protein Polymers Active in the Cytoplasm of Bacterial and Archaeal Cells* (pp. 161–211). Springer International Publishing. https://doi.org/10.1007/978-3-319-53047-5_6
68. Kennedy, I. R., Choudhury, A. T. M. A., & Kecskés, M. L. (2004). Non-symbiotic bacterial diazotrophs in crop-farming systems: Can their potential for plant growth promotion be better exploited? *Soil Biology and Biochemistry*, *36*(8), 1229–1244. https://doi.org/10.1016/j.soilbio.2004.04.006
69. Kordrostami, M., Rabiei, B., & Ebadi, A. A. (2019). Oxidative Stress in Plants: Production, Metabolism, and Biological Roles of Reactive Oxygen Species. In *Handbook of Plant and Crop Stress, Fourth Edition* (4th ed.). CRC Press.
70. Kovachev, S. (2018). Defence factors of vaginal lactobacilli. *Critical Reviews in Microbiology*, *44*(1), 31–39. https://doi.org/10.1080/1040841X.2017.1306688
71. Lang, M., Carvalho, A., Baharoglu, Z., & Mazel, D. (2023). Aminoglycoside uptake, stress, and potentiation in Gram-negative bacteria: New therapies with old molecules. *Microbiology and Molecular Biology Reviews*, *87*(4), e00036-22. https://doi.org/10.1128/mmbr.00036-22
72. Larkins, B. A. (1985). Polyribosomes. In H.-F. Linskens & J. F. Jackson (Eds.), *Cell Components* (pp. 331–352). Springer. https://doi.org/10.1007/978-3-642-82587-3_16
73. Leak, D. J., & Dalton, H. (1986). Growth yields of methanotrophs. *Applied Microbiology and Biotechnology*, *23*(6), 470–476. https://doi.org/10.1007/BF02346062
74. LeBlanc, J. G., Burgess, C., Sesma, F., de Giori, G. S., & van Sinderen, D. (2005). Ingestion of Milk Fermented by Genetically Modified *Lactococcus lactis* Improves the Riboflavin Status of Deficient Rats. *Journal of Dairy Science*, *88*(10), 3435–3442. https://doi.org/10.3168/jds.S0022-0302(05)73027-7
75. Lemire, B. D., & Oyedotun, K. S. (2002). The *Saccharomyces cerevisiae*

mitochondrial succinate:ubiquinone oxidoreductase. *Biochimica et Biophysica Acta (BBA) - Bioenergetics*, *1553*(1), 102–116. https://doi.org/10.1016/S0005-2728(01)00229-8

76. Lenz, C. A., Hew Ferstl, C. M., & Vogel, R. F. (2010). Sub-lethal stress effects on virulence gene expression in *Enterococcus faecalis*. *Food Microbiology*, *27*(3), 317–326. https://doi.org/10.1016/j.fm.2009.11.008
77. Li, G., Walker, M. J., & De Oliveira, D. M. P. (2023). Vancomycin Resistance in Enterococcus and Staphylococcus aureus. *Microorganisms*, *11*(1), Article 1. https://doi.org/10.3390/microorganisms11010024
78. Linderman, R. G. (2015). Vesicular-Arbuscular Mycorrhizae and Soil Microbial Interactions. In G. J. Bethlenfalvay & R. G. Linderman (Eds.), *ASA Special Publications* (pp. 45–70). American Society of Agronomy, Crop Science Society of America, and Soil Science Society of America. https://doi.org/10.2134/asaspecpub54.c3
79. Liu, W., Pang, H., Zhang, H., & Cai, Y. (2014). Biodiversity of Lactic Acid Bacteria. In H. Zhang & Y. Cai (Eds.), *Lactic Acid Bacteria* (pp. 103–203). Springer Netherlands. https://doi.org/10.1007/978-94-017-8841-0_2
80. Lock, M. A. (1994). Dynamics of Particulate and Dissolved Organic Matter over the Substratum of Water Bodies. In *The Biology of Particles in Aquatic Systems, Second Edition* (2nd ed.). CRC Press.
81. Lockhart, D. J., & Winzeler, E. A. (2000). Genomics, gene expression and DNA arrays. *Nature*, *405*(6788), 827–836. https://doi.org/10.1038/35015701
82. Łozińska, N., Maldonado-Valderrama, J., Del Castillo-Santaella, T., Zhou, Y., Martysiak-Żurowska, D., Lu, Y., & Jungnickel, C. (2024). Bile conjugation and its effect on in vitro lipolysis of emulsions. *Food Research International*, *184*, 114255. https://doi.org/10.1016/j.foodres.2024.114255
83. Lu, Q., He, Z. L., Graetz, D. A., Stoffella, P. J., & Yang, X. (2010). Phytoremediation to remove nutrients and improve eutrophic stormwaters using water lettuce (Pistia stratiotes L.). *Environmental Science and Pollution Research*, *17*(1), 84–96. https://doi.org/10.1007/s11356-008-0094-0
84. Lubitz, W., Chrysina, M., & Cox, N. (2019). Water oxidation in photosystem II. *Photosynthesis Research*, *142*(1), 105–125. https://doi.org/10.1007/s11120-019-00648-3
85. Lyu, D., Msimbira, L. A., Nazari, M., Antar, M., Pagé, A., Shah, A., Monjezi, N., Zajonc, J., Tanney, C. A. S., Backer, R., & Smith, D. L. (2021). The Coevolution of Plants and Microbes Underpins Sustainable

Agriculture. *Microorganisms*, *9*(5), 1036. https://doi.org/10.3390/microorganisms9051036

86. Mailloux, R. J. (2020). Protein S-glutathionylation reactions as a global inhibitor of cell metabolism for the desensitization of hydrogen peroxide signals. *Redox Biology*, *32*, 101472. https://doi.org/10.1016/j.redox.2020.101472
87. Maliva, R. (2021). Intergovernmental Panel on Climate Change and Global Climate Change Projections. In R. Maliva (Ed.), *Climate Change and Groundwater: Planning and Adaptations for a Changing and Uncertain Future: WSP Methods in Water Resources Evaluation Series No. 6* (pp. 71–88). Springer International Publishing. https://doi.org/10.1007/978-3-030-66813-6_4
88. Marchant, R., Banat, I. M., Rahman, T. J., & Berzano, M. (2002). The frequency and characteristics of highly thermophilic bacteria in cool soil environments. *Environmental Microbiology*, *4*(10), 595–602. https://doi.org/10.1046/j.1462-2920.2002.00344.x
89. Marschner, B., & Kalbitz, K. (2003). Controls of bioavailability and biodegradability of dissolved organic matter in soils. *Geoderma*, *113*(3), 211–235. https://doi.org/10.1016/S0016-7061(02)00362-2
90. Martinac, B., Saimi, Y., & Kung, C. (2008). Ion Channels in Microbes. *Physiological Reviews*, *88*(4), 1449–1490. https://doi.org/10.1152/physrev.00005.2008
91. Marulanda, A., Porcel, R., Barea, J. M., & Azcón, R. (2007). Drought Tolerance and Antioxidant Activities in Lavender Plants Colonized by Native Drought-tolerant or Drought-sensitive Glomus Species. *Microbial Ecology*, *54*(3), 543–552. https://doi.org/10.1007/s00248-007-9237-y
92. Monk, B. C., & Perlin, D. S. (1994). Fungal Plasma Membrane Proton Pumps as Promising New Antifungal Targets. *Critical Reviews in Microbiology*, *20*(3), 209–223. https://doi.org/10.3109/10408419409114555
93. Moyes, R. B., Reynolds, J., & Breakwell, D. P. (2009). Differential Staining of Bacteria: Gram Stain. *Current Protocols in Microbiology*, *15*(1), A.3C.1-A.3C.8. https://doi.org/10.1002/9780471729259.mca03cs15
94. Nadeem, S. M., Ahmad, M., Zahir, Z. A., Javaid, A., & Ashraf, M. (2014). The role of mycorrhizae and plant growth promoting rhizobacteria (PGPR) in improving crop productivity under stressful environments. *Biotechnology Advances*, *32*(2), 429–448. https://doi.org/10.1016/j.biotechadv.2013.12.005
95. Neeraja, C., Anil, K., Purushotham, P., Suma, K., Sarma, P., Moerschbacher, B. M., & Podile, A. R. (2010). Biotechnological approaches to develop

bacterial chitinases as a bioshield against fungal diseases of plants. *Critical Reviews in Biotechnology*, *30*(3), 231–241. https://doi.org/10.3109/07388551.2010.487258

96. Neher, D. A. (2010). Ecology of Plant and Free-Living Nematodes in Natural and Agricultural Soil. *Annual Review of Phytopathology*, *48*(Volume 48, 2010), 371–394. https://doi.org/10.1146/annurev-phyto-073009-114439
97. Nielsen, D. R., Th. Van Genuchten, M., & Biggar, J. W. (1986). Water flow and solute transport processes in the unsaturated zone. *Water Resources Research*, *22*(9S), 89S-108S. https://doi.org/10.1029/WR022i09Sp0089S
98. Oakley, B. R. (2000). An abundance of tubulins. *Trends in Cell Biology*, *10*(12), 537–542. https://doi.org/10.1016/S0962-8924(00)01857-2
99. Orlova, A., Garner, E. C., Galkin, V. E., Heuser, J., Mullins, R. D., & Egelman, E. H. (2007). The structure of bacterial ParM filaments. *Nature Structural & Molecular Biology*, *14*(10), 921–926. https://doi.org/10.1038/nsmb1300
100. Ourisson, G., Rohmer, M., & Poralla, K. (1987). Prokaryotic Hopanoids and other Polyterpenoid Sterol Surrogates. *Annual Review of Microbiology*, *41*(Volume 41,), 301–333. https://doi.org/10.1146/annurev.mi.41.100187.001505
101. Palmgren, M., & Morsomme, P. (2019). The plasma membrane H+-ATPase, a simple polypeptide with a long history. *Yeast*, *36*(4), 201–210. https://doi.org/10.1002/yea.3365
102. Pang, B., McFaline, J. L., Burgis, N. E., Dong, M., Taghizadeh, K., Sullivan, M. R., Elmquist, C. E., Cunningham, R. P., & Dedon, P. C. (2012). Defects in purine nucleotide metabolism lead to substantial incorporation of xanthine and hypoxanthine into DNA and RNA. *Proceedings of the National Academy of Sciences*, *109*(7), 2319–2324. https://doi.org/10.1073/pnas.1118455109
103. Panse, V. G., & Johnson, A. W. (2010). Maturation of eukaryotic ribosomes: Acquisition of functionality. *Trends in Biochemical Sciences*, *35*(5), 260–266. https://doi.org/10.1016/j.tibs.2010.01.001
104. Paul, C., Filippidou, S., Jamil, I., Kooli, W., House, G. L., Estoppey, A., Hayoz, M., Junier, T., Palmieri, F., Wunderlin, T., Lehmann, A., Bindschedler, S., Vennemann, T., Chain, P. S. G., & Junier, P. (2019). Chapter Three—Bacterial spores, from ecology to biotechnology. In G. M. Gadd & S. Sariaslani (Eds.), *Advances in Applied Microbiology* (Vol. 106, pp. 79–111). Academic Press. https://doi.org/10.1016/bs.aambs.2018.10.002
105. Pavlova, M. D., Asaturova, A. M., & Kozitsyn, A. E. (2022). Bacterial

Cell Shape: Some Features of Ultrastructure, Evolution, and Ecology. *Biology Bulletin Reviews*, *12*(3), 254–265. https://doi.org/10.1134/S2079086422030070

106. Plavec, T. V., & Berlec, A. (2020). Safety Aspects of Genetically Modified Lactic Acid Bacteria. *Microorganisms*, *8*(2), 297. https://doi.org/10.3390/microorganisms8020297
107. Pócsi, I. (2011). Toxic Metal/Metalloid Tolerance in Fungi—A Biotechnology-Oriented Approach. In G. Banfalvi (Ed.), *Cellular Effects of Heavy Metals* (pp. 31–58). Springer Netherlands. https://doi.org/10.1007/978-94-007-0428-2_2
108. Rashmi, D., Zanan, R., John, S., Khandagale, K., & Nadaf, A. (2018). γ-Aminobutyric Acid (GABA): Biosynthesis, Role, Commercial Production, and Applications. In *Studies in Natural Products Chemistry* (Vol. 57, pp. 413–452). Elsevier. https://doi.org/10.1016/B978-0-444-64057-4.00013-2
109. Rautio, M., Mariash, H., & Forsström, L. (2011). Seasonal shifts between autochthonous and allochthonous carbon contributions to zooplankton diets in a subarctic lake. *Limnology and Oceanography*, *56*(4), 1513–1524. https://doi.org/10.4319/lo.2011.56.4.1513
110. Razin, S. (1981). The Mycoplasma Membrane. In *Organization of Prokaryotic Cell Membranes*. CRC Press.
111. Rieder, S. E., & Emr, S. D. (2000). Isolation of Subcellular Fractions from the Yeast Saccharomyces cerevisiae. *Current Protocols in Cell Biology*, *8*(1), 3.8.1-3.8.68. https://doi.org/10.1002/0471143030.cb0308s08
112. Rivett, M. O., Feenstra, S., & Cherry, J. A. (2001). A controlled field experiment on groundwater contamination by a multicomponent DNAPL: Creation of the emplaced-source and overview of dissolved plume development. *Journal of Contaminant Hydrology*, *49*(1), 111–149. https://doi.org/10.1016/S0169-7722(00)00191-1
113. Rodrigues, J. L. M., Aiello, M. R., Urbance, J. W., Tsoi, T. V., & Tiedje, J. M. (2002). Use of both 16S rRNA and engineered functional genes with real-time PCR to quantify an engineered, PCB-degrading *Rhodococcus* in soil. *Journal of Microbiological Methods*, *51*(2), 181–189. https://doi.org/10.1016/S0167-7012(02)00067-2
114. Saleem, M., Nawaz, F., Hussain, M. B., & Ikram, R. M. (2021). Comparative Effects of Individual and Consortia Plant Growth Promoting Bacteria on Physiological and Enzymatic Mechanisms to Confer Drought Tolerance in Maize (Zea mays L.). *Journal of Soil Science and Plant Nutrition*, *21*(4), 3461–3476. https://doi.org/10.1007/s42729-021-00620-y

115. Sánchez-Pascuala, A., De Lorenzo, V., & Nikel, P. I. (2017). Refactoring the Embden–Meyerhof–Parnas Pathway as a Whole of Portable GlucoBricks for Implantation of Glycolytic Modules in Gram-Negative Bacteria. *ACS Synthetic Biology*, *6*(5), 793–805. https://doi.org/10.1021/acssynbio.6b00230

116. Schmutzhard, E., & Helbok, R. (2014). Chapter 96—Rickettsiae, protozoa, and opisthokonta/metazoa. In J. Biller & J. M. Ferro (Eds.), *Handbook of Clinical Neurology* (Vol. 121, pp. 1403–1443). Elsevier. https://doi.org/10.1016/B978-0-7020-4088-7.00096-1

117. Selvameenal, L., Radhakrishnan, M., & Balagurunathan, R. (2009). Antibiotic Pigment from Desert Soil Actinomycetes; Biological Activity, Purification and Chemical Screening. *Indian Journal of Pharmaceutical Sciences*, *71*(5), 499–504. https://doi.org/10.4103/0250-474X.58174

118. Seshasayee, A. S. N., Sivaraman, K., & Luscombe, N. M. (2011). An Overview of Prokaryotic Transcription Factors. In T. R. Hughes (Ed.), *A Handbook of Transcription Factors* (pp. 7–23). Springer Netherlands. https://doi.org/10.1007/978-90-481-9069-0_2

119. Seydel, U., Wiese, A., Schromm, A. B., & Brandenburg, K. (1999). A Biophysical View on the Function and Activity of Endotoxins. In *Endotoxin in Health and Disease*. CRC Press.

120. Sheen, J. (1994). Feedback control of gene expression. *Photosynthesis Research*, *39*(3), 427–438. https://doi.org/10.1007/BF00014596

121. Shukla, N., Awasthi, R. P., Rawat, L., & Kumar, J. (2015). Seed biopriming with drought tolerant isolates of *Trichoderma harzianum* promote growth and drought tolerance in *Triticum aestivum*: *Trichoderma* mediated drought responses in bioprimed wheat plants. *Annals of Applied Biology*, *166*(2), 171–182. https://doi.org/10.1111/aab.12160

122. Simons, A., Alhanout, K., & Duval, R. E. (2020). Bacteriocins, Antimicrobial Peptides from Bacterial Origin: Overview of Their Biology and Their Impact against Multidrug-Resistant Bacteria. *Microorganisms*, *8*(5), Article 5. https://doi.org/10.3390/microorganisms8050639

123. Sing, H., Löffler, F. E., & Fathepure, B. Z. (2004). Aerobic biodegradation of vinyl chloride by a highly enriched mixed culture. *Biodegradation*, *15*(3), 197–204. https://doi.org/10.1023/B:BIOD.0000026539.55941.73

124. Singh, G., & Mukerji, K. G. (2006). Root Exudates as Determinant of Rhizospheric Microbial Biodiversity. In K. G. Mukerji, C. Manoharachary, & J. Singh (Eds.), *Microbial Activity in the Rhizoshere* (Vol. 7, pp. 39–53). Springer-Verlag. https://doi.org/10.1007/3-540-29420-1_3

125. Singh, S. K., Wu, X., Shao, C., & Zhang, H. (2022). Microbial

enhancement of plant nutrient acquisition. *Stress Biology*, *2*(1), 3. https://doi.org/10.1007/s44154-021-00027-w

126. Sommer, U. (2023). Life Forms of Aquatic Organisms. In U. Sommer (Ed.), *Freshwater and Marine Ecology* (pp. 53–113). Springer International Publishing. https://doi.org/10.1007/978-3-031-42459-5_3
127. Song, Y., Wilson, A. J., Zhang, X.-C., Thoms, D., Sohrabi, R., Song, S., Geissmann, Q., Liu, Y., Walgren, L., He, S. Y., & Haney, C. H. (2021). FERONIA restricts Pseudomonas in the rhizosphere microbiome via regulation of reactive oxygen species. *Nature Plants*, *7*(5), 644–654. https://doi.org/10.1038/s41477-021-00914-0
128. Spirin, A. S. (2002). Ribosome as a molecular machine. *FEBS Letters*, *514*(1), 2–10. https://doi.org/10.1016/S0014-5793(02)02309-8
129. Steinbacher, S., Bass, R., Strop, P., & Rees, D. C. (2007). Structures of the Prokaryotic Mechanosensitive Channels MscL and MscS. In *Current Topics in Membranes* (Vol. 58, pp. 1–24). Elsevier. https://doi.org/10.1016/S1063-5823(06)58001-9
130. Summer, E. J. (2009). Preparation of a Phage DNA Fragment Library for Whole Genome Shotgun Sequencing. In M. R. J. Clokie & A. M. Kropinski (Eds.), *Bacteriophages: Methods and Protocols, Volume 2 Molecular and Applied Aspects* (pp. 27–46). Humana Press. https://doi.org/10.1007/978-1-60327-565-1_4
131. Swain, M. R., Anandharaj, M., Ray, R. C., & Parveen Rani, R. (2014). Fermented Fruits and Vegetables of Asia: A Potential Source of Probiotics. *Biotechnology Research International*, *2014*(1), 250424. https://doi.org/10.1155/2014/250424
132. Taillefer, M., & Sparling, R. (2016). Glycolysis as the Central Core of Fermentation. In R. Hatti-Kaul, G. Mamo, & B. Mattiasson (Eds.), *Anaerobes in Biotechnology* (pp. 55–77). Springer International Publishing. https://doi.org/10.1007/10_2015_5003
133. Takahashi, S., Abe, K., & Kera, Y. (2015). Bacterial d-amino acid oxidases: Recent findings and future perspectives. *Bioengineered*, *6*(4), 237–241. https://doi.org/10.1080/21655979.2015.1052917
134. Tan, J. K., McKenzie, C., Mariño, E., Macia, L., & Mackay, C. R. (2017). Metabolite-Sensing G Protein–Coupled Receptors—Facilitators of Diet-Related Immune Regulation. *Annual Review of Immunology*, *35*(1), 371–402. https://doi.org/10.1146/annurev-immunol-051116-052235
135. Taylor, A. J., & Kelly, D. J. (2019). The function, biogenesis and regulation of the electron transport chains in *Campylobacter jejuni*. In R.

K. Poole (Ed.), *Advances in Microbial Physiology* (Vol. 74, pp. 239–329). Academic Press. https://doi.org/10.1016/bs.ampbs.2019.02.003

136. Ushio, K., Sawatani, M., & Nakata, Y. (1991). Effect of altered sterol composition on the salt tolerance of *Zygosaccharomyces rouxii*. *Journal of Fermentation and Bioengineering*, *71*(6), 390–396. https://doi.org/10.1016/0922-338X(91)90249-G
137. Van Bergeijk, S. A., Van der Zee, C., & Stal, L. J. (2003). Uptake and excretion of dimethylsulphoniopropionate is driven by salinity changes in the marine benthic diatom Cylindrotheca closterium. *European Journal of Phycology*, *38*(4), 341–349. https://doi.org/10.1080/09670260310001612600
138. Van Colen, C., Underwood, G. J. C., Serôdio, J., & Paterson, D. M. (2014). Ecology of intertidal microbial biofilms: Mechanisms, patterns and future research needs. *Journal of Sea Research*, *92*, 2–5. https://doi.org/10.1016/j.seares.2014.07.003
139. Vega, M. G. de la, Cejudo, F. J., & Paneque, A. (1991). Purification and properties of an extracellular invertase from *Azotobacter chroococcum*. *Enzyme and Microbial Technology*, *13*(3), 267–271. https://doi.org/10.1016/0141-0229(91)90140-6
140. Velázquez, E., Carro, L., Flores-Félix, J. D., Menéndez, E., Ramírez-Bahena, M.-H., & Peix, A. (2019). Bacteria-Inducing Legume Nodules Involved in the Improvement of Plant Growth, Health and Nutrition. In V. Kumar, R. Prasad, M. Kumar, & D. K. Choudhary (Eds.), *Microbiome in Plant Health and Disease: Challenges and Opportunities* (pp. 79–104). Springer. https://doi.org/10.1007/978-981-13-8495-0_4
141. Wagstaff, J. M., Tsim, M., Oliva, M. A., García-Sanchez, A., Kureisaite-Ciziene, D., Andreu, J. M., & Löwe, J. (2017). A Polymerization-Associated Structural Switch in FtsZ That Enables Treadmilling of Model Filaments. *mBio*, *8*(3), 10.1128/mbio.00254-17. https://doi.org/10.1128/mbio.00254-17
142. Wang, Y., Li, J., & Liu, A. (2017). Oxygen activation by mononuclear nonheme iron dioxygenases involved in the degradation of aromatics. *JBIC Journal of Biological Inorganic Chemistry*, *22*(2), 395–405. https://doi.org/10.1007/s00775-017-1436-5
143. Weiss, D. S. (2004). Bacterial cell division and the septal ring. *Molecular Microbiology*, *54*(3), 588–597. https://doi.org/10.1111/j.1365-2958.2004.04283.x
144. Widmer, F. (2007). Assessing Effects of Transgenic Crops on Soil Microbial Communities. In A. Fiechter & C. Sautter (Eds.), *Green Gene*

Technology (Vol. 107, pp. 207–234). Springer Berlin Heidelberg. https://doi.org/10.1007/10_2007_047

145. Wirth, P. J., & Romano, A. (1995). Staining methods in gel electrophoresis, including the use of multiple detection methods. *Journal of Chromatography A*, *698*(1), 123–143. https://doi.org/10.1016/0021-9673(94)00879-E
146. Wu, W., & Li, H. (2018). Metabolites of Lactic Acid Bacteria. In W. Chen & A. Narbad, *Lactic Acid Bacteria in Foodborne Hazards Reduction* (pp. 87–113). Springer Singapore. https://doi.org/10.1007/978-981-13-1559-6_4
147. Xiong, C., Feng, S., Qiao, Y., Guo, Z., & Gu, G. (2018). Synthesis and Immunological Studies of Oligosaccharides that Consist of the Repeating Unit of Streptococcus pneumoniae Serotype 3 Capsular Polysaccharide. *Chemistry – A European Journal*, *24*(32), 8205–8216. https://doi.org/10.1002/chem.201800754
148. Yang, Y., & Sauve, A. A. (2016). NAD+ metabolism: Bioenergetics, signaling and manipulation for therapy. *Biochimica et Biophysica Acta (BBA) - Proteins and Proteomics*, *1864*(12), 1787–1800. https://doi.org/10.1016/j.bbapap.2016.06.014
149. Yaquoob, P., Tricon, S., Burdge, G. C., & Calder, P. C. (2006). Conjugated linoleic acids (CLAs) and health. In *Improving the Fat Content of Foods* (pp. 182–209). Elsevier. https://doi.org/10.1533/9781845691073.1.182
150. Yee, B., Lafi, F. F., Oakley, B., Staley, J. T., & Fuerst, J. A. (2007). A canonical FtsZ protein in Verrucomicrobium spinosum, a member of the Bacterial phylum Verrucomicrobia that also includes tubulin-producing Prosthecobacter species. *BMC Evolutionary Biology*, *7*(1), 37. https://doi.org/10.1186/1471-2148-7-37
151. Zhang, N., Wang, D., Liu, Y., Li, S., Shen, Q., & Zhang, R. (2014). Effects of different plant root exudates and their organic acid components on chemotaxis, biofilm formation and colonization by beneficial rhizosphere-associated bacterial strains. *Plant and Soil*, *374*(1–2), 689–700. https://doi.org/10.1007/s11104-013-1915-6
152. Zohri, A.-N. A., Kassim, R. M. F., & Hassan, S. H. A. (2024). Methylene blue as an exogenous electron mediator on bioelectricity from molasses using Meyerozyma guilliermondii as biocatalyst. *Biomass Conversion and Biorefinery*, *14*(5), 6649–6657. https://doi.org/10.1007/s13399-022-03016-9

INDEX

C

D

E

F

G

H

O

P

Q

R

S

FSC
www.fsc.org
MIX
Papier | Fördert
gute Waldnutzung
FSC® C083411

Kolibri 360 GmbH
Ferdinand-Jühlke-Straße 7
99095 Erfurt, Deutschland
produktsicherheit@kolibri360.de